Biological Techniques Series

Alexander Hollaender, *Editor*

BIOLOGICAL
MICROIRRADIATION
Classical and Laser Sources

Michael W. Berns

University of California
Irvine, California

Prentice-Hall, Inc., Englewood Cliffs, N. J.

Library of Congress Cataloging in Publication Data

Berns, Michael W
  Biological microirradiation.

  (Biological techniques series)
  Bibliography: p.
  1. Radiobiology—Technique. 2. Irradiation.
I. Title. DNLM: 1. Cells—Radiation effects.
2. Lasers. 3. Radiobiology. QH652 B531b 1974
QH652.B46   574.1'915   73-18329
ISBN 0-13-077032-9

PRENTICE-HALL INTERNATIONAL, INC., *London*
PRENTICE-HALL OF AUSTRALIA, PTY. LTD., *Sydney*
PRENTICE-HALL OF CANADA, LTD., *Toronto*
PRENTICE-HALL OF INDIA PRIVATE LIMITED, *New Delhi*
PRENTICE-HALL OF JAPAN, INC., *Toyko*

10 9 8 7 6 5 4 3 2 1

Printed in the United States of America

CONTENTS

v

PREFACE

The motivating factor in writing this book was the general misconception that exists with respect to microbeams and their application. It is not uncommon for investigators to be openly pessimistic about the use of microbeams in the quest for answers to their problems. On the other hand, it is not uncommon to find investigators who believe that the microbeam can answer all of their questions. Both conceptions are extremes and result from a general misunderstanding about the technique. This can stem from a poor understanding of radiological principles, naiveté with respect to the complexity *and* simplicity of the apparatus, a lack of familiarity with the past procedures, or even a poor understanding of the biological system being investigated.

This book is an attempt to place all of these factors in perspective. The initial chapters present some basic principles of radiation biology, equipment design and operation, and inherent limitations of the technique. Subsequent chapters treat biological microirradiation in both a historical and current context. A wide variety of experiments are discussed in order to provide the reader with enough background so that he/she can judge for himself/herself whether or not the microbeam approach stands a reasonable chance of being successful.

This book is designed so as to provide all the information necessary for the investigator who is considering using the approach in the laboratory. In addition it provides all the basic principles involved in microirradiation as well as a rather extensive review of the literature. This book should serve to introduce the student and the researcher to a biological approach that has been used since 1912 and which was, and still is, contributing to scientific progress in a large number of biological disciplines.

I am deeply indebted to those colleagues who personally encouraged and motivated me towards an academic career. Specifically, William T. Keeton and Lowell D. Uhler provided me with the impetus and support necessary to establish and pursue my goals. I am particularly grateful to Donald E. Rounds for providing an environment in which I could learn cell culture procedures and further develop my interests in employing lasers for partial cell irradiation. Particular thanks are given to Drs. Marcel Bessis, Giuiliana Moreno, and Christian Salet for the generous supply of photographs and illustrations that are contained in this manuscript. In addition I would like to thank Drs. Raymond Zirkle, Henry Harris, Robert Perry, Philip Dendy, Norman Saks, Joe Griffin, R. Storb, and J. Daniel for kindly providing original copies of illustrations.

MICHAEL W. BERNS
Irvine, California

Introduction

ONE

In 1912 the Russian Tchakhotine built the first microirradiation device for the purpose of destroying small regions of single cells, organisms, or embryos. From the years 1912 to 1959 this prolific investigator employed his ultraviolet microbeam to irradiate organisms such as sea urchin eggs, bivalve molluscs, numerous types of protozoans, spermatozoa, algae, etc. The types of biological problems studied varied from basic developmental biology (fertilization, determination, and parthenogenesis) to problems of cell physiology (such as contractility and cell motility) and cell divison. It is fair to say that if there ever were a founder of a science or a particular scientific method of approach, Tchakhotine would have to be designated the father of microirradiation.

From his early work employing ultraviolet microirradiation for the study of basic biological processes, several sub-applications were derived. Microirradiation, for example, has been employed using numerous radiation sources, generally with three purposes in mind: (1) to elucidate cell or organism function by altering or destroying a partial region of the target system (this was the major approach as originally employed by Tchakhotine); (2) for defining the radiosensitivity of various subcellular regions and structures (an approach primarily of a radiobiological nature); and (3), strictly as an

1

analytical tool for reading-out quantitative and qualitative cellular information.

When one consults the four major biological reviews of microirradiation (Zirkle, 1957; Moreno, Lutz, and Bessis, 1969; Smith, 1964; and Berns and Salet, 1972), it becomes obvious that in terms of published research and, presumably, interest, the vast majority of investigators have been employing microirradiation to study biological function rather than radiobiology, or as an analytical biological tool. However, this statement may be somewhat misleading, because literally hundreds of studies have been performed employing microspectrophotometry and microfluorimetry with focused ultraviolet radiation. In these studies ultraviolet microirradiation is employed in an analytical way to generate quantitative and qualitative information about cells and cell structures. In these studies the cells are often dead, therefore, cell survival for even a short time is not of interest (Caspersson, 1968; Glubrecht, 1958, 1960, 1963).

Similarly, Glick (1969) has employed a focused laser to read-out the low levels of elements within single cells by flash spectroscopy. By analyzing the spectrum of the incandescent light emitted from a vaporized sample, it has been possible to make elemental determinations as small as 10^{-9} g. The fact that the people who have written the reviews on microbeams (myself included) employ microirradiation to study biological function is probably why *analytical* microirradiation has been generally ignored. In addition, the analytical microirradiation systems, such as a microspectrophotometer, have evolved to such a refined level of sophistication and automation, that an investigator need only place his specimen under the system and make the appropriate measurements. Reviewing the hundreds of studies employing these techniques would be quite impossible. Consequently, in this volume I will follow the convention of treating biological microirradiation as an approach to study biological function by *partial irradiation* and also as a method to study radiosensitivity of biological systems, while at least acknowledging the existence of other microirradiation-like systems.

I should like to address myself for a moment to terminology. As the title of this book implies, the subject is really *partial irradiation*. The use of the phrases *partial cell irradiation* (PCI) and *microbeams* has been purposely avoided. Indeed numerous studies involving partial irradiation of whole multicellular organisms, embryos, and tissues have been performed. Truly, these are not studies in partial *cell* irradiation. Similarly, numerous partial irradiation devices are not microbeams. Large macrobeams that employ partial shielding of the target are not microbeams. The microsource of polonium alpha particles (Munro, 1957) is not a microbeam. Therefore, the more general, all-inclusive phrases *partial irradiation of biological systems* and *biological microirradiation* are employed extensively in this book.

In writing a volume on a biological technique, one can devote much space to a description of methodology. However, for such a work to be useful to the large number of diverse scientists, one must deal with the kinds of problems that one might encounter and the types of questions that can be answered. I have devoted approximately half of this book to a description of methodology, equipment, and basic radiation biology, and the remaining half to a discussion of the biological questions and answers approached by the technique. As a result, numerous studies are not mentioned at all, some studies are discussed only briefly, and others are discussed in great detail. I have attempted to select a very diverse group of investigations for discussion. By doing this, it is my hope that any biologist picking up this book can rapidly determine whether or not his questions can be answered.

Finally, as a biologist, I find it particularly difficult to write a volume dealing only with a technique. Therefore, I have attempted throughout the second half of this volume to deal at length with those microirradiation studies that have contributed to the solution, or partial understanding, of significant biological questions.

Instrumentation

TWO

There are as many different microbeam devices as there are investigators using them. As a general rule each microbeamist designs and builds an instrument suited to his or her particular needs. These needs are reflected in the type of radiation employed, the nature of the material irradiated, and the kinds of questions being asked. A discussion of the variety of devices would take an entire book. However, the various methods of PCI can be lumped into four general categories, of which all employ a radiation source external to the cell or organism.

A. MACROBEAM FOCUSED TO A MICROBEAM

One of the most widely employed methods of microbeam irradiation is to take a polychromatic source of UV light, separate the desired wavelength, and focus it to a small spot within the target specimen. This was the first method of PCI reported by the Russian Tchakhotine in 1912 and subsequently refined and described by him in 1935. This system is diagrammed in Fig. 2-1. Light from a magnesium spark is collected by a quartz lens and directed into two quartz prisms arranged to refract the 2800 Å light down the optical

4

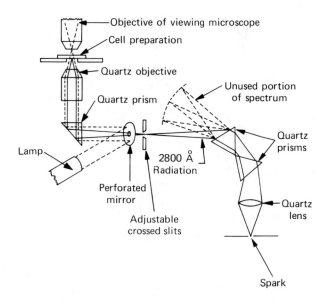

Fig. 2-1. Tchakhotine's 2800 Å microbeam device. The 2800 Å light is designated by an unbroken line; the unused portion of the spark spectrum and the visible viewing light is designated by a broken line. (From Zirkle, 1957.)

system. The radiation passes through a rectangular aperture formed by two adjustable slits and through a perforated mirror. It is next reflected by a quartz prism into the quartz objective that focuses the beam to a small spot within the cell. The cell is viewed with a regular compound microscope using visible substage transmitted illumination that is reflected off the perforated mirror, quartz prism, and through the quartz objective, which acts as the condenser. Aiming of the microbeam is accomplished by focusing the UV light into a drop of fluorescent dye (fluorescein) that is placed on a microscope slide. The tip of a movable pointer in the ocular is located at the center of the *hot spot* of fluorescence. The fluorescent specimen is removed and replaced with the target cell, which is moved until the part selected for irradiation is located under the tip of the pointer.

Tchakhotine's method of focused UV microirradiation was the primary microbeam technique until 1954 when Uretz and co-workers at the University of Chicago developed the reflecting objective. Their system is diagrammed in Fig. 2-2. The reflecting objective is

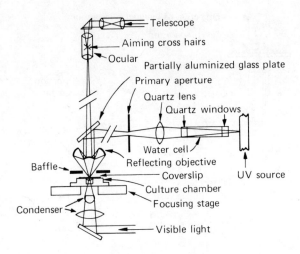

Fig. 2-2.   Reflecting objective ultraviolet microbeam. (From Zirkle, 1957.)

utilized both for focusing the incident radiation and as the observation objective. A standard substage condenser and light source provide illumination. The UV light is collected by a quartz lens after traveling through a water cell and is directed into an adjustable aperture. It then passes through the aperture and is reflected into the objective by a partially aluminized mirror (coated to transmit visible light; e.g., the specimen image). The UV image of the aperture (microimage) is focused onto the same plane as the viewing microscope by adjusting the distance between the aperture and the objective. The microspot of UV light can be located directly under the cross hair in the ocular by moving the microaperture in a plane perpendicular to the beam axis. The specimen is placed under the microscope and the area to be irradiated is moved under the crosshair. The opinion expressed by Zirkle in his 1957 review was that the reflecting objective provides an apparatus that "is simpler, much less expensive, more accurately aimed, more flexible in use, wider in application, and much less demanding of operational labor than Tchakhotine's."[1]   However, it must be pointed out that the reflecting objective does have a small numerical aperture; it is difficult, but possible, to incorporate phase into the system; and it

[1] R. E. Zirkle, "Partial Cell Irradiation," *Adv. Biol. Med. Phys.*, 5 (1957), 122.

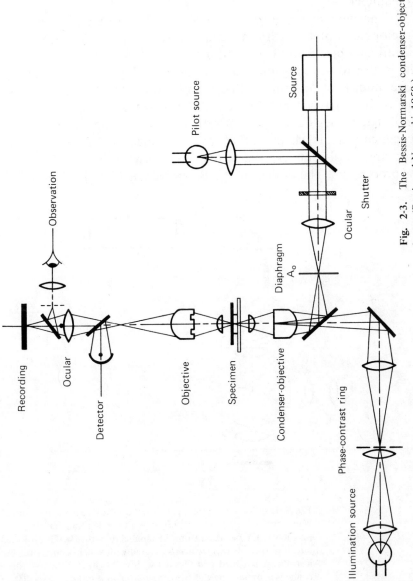

Fig. 2-3. The Bessis-Normarski condenser-objective UV microbeam. (Bessis and Nomarski, 1959.)

Pilot source

Source

Observation

Shutter

Ocular

Diaphragm
A_o

Recording

Ocular

Objective

Specimen

Condenser-objective

Detector

Phase-contrast ring

Illumination source

cannot focus light to a spot less than several micrometers (2-10μm) in diameter.

The next major advance in UV microbeam irradiation was the development of the achromatically corrected (2310-7000 Å) Zeiss Ultrafluar objective. The development of this objective permitted the return to transmission phase objectives of high magnification and large numerical aperture. Bessis and Nomarski (1959) developed one of the first systems (Fig. 2-3). An image of the aperture (A_o is projected through a substage ultrafluar objective-condenser onto the specimen. It is claimed that by changing the diameter of the aperture, focal spots varying in diameter between 0.2 and 10μm can be produced. The design of the Bessis-Nomarski system returns to

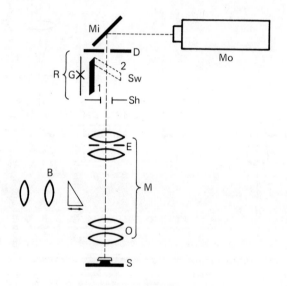

Fig. 2-4. The UV optical path of the reflex camera microbeam. When the swinging mirror (*Sw*) of the reflex camera (*R*) is in position 1, UV radiation from the monochromator (*Mo*) is projected onto the specimen (*S*) which is covered with a quartz coverslip. The instrument is aimed and focused with visible light from a substage condenser by moving the swinging mirror into position 2 so that an image of the specimen is projected on the ground glass screen (*G*) and brought into coincidence with the cross. First surface aluminized mirror. *Mi*; field diaphragm, *D* with an aperture which defines the size of the microbeam; shutter, *Sh*; microscope, *M*; ultrafluar eyepiece, *E*; ultra objective, *O*; binocular eyepieces with slide-in prism, *B*. (From Rustad, 1968).

the original two microscope system of Tchakhotine; one for viewing and one for focusing. The ultrafluar objective has permitted construction of several different microbeam systems that utilize the same objective both for viewing and focusing. One of the most recent systems is diagrammed in Fig. 2-4. This unique system utilized a reflex camera in conjunction with a microscope equipped with ultrafluar optics.

In addition to the development of the ultafluar objective, Montgomery and Hundly (1961) developed a unique UV microbeam utilizing *flying-spot* television microscopy (Fig. 2-5). The image of the UV-emitting scanning tube is focused on the object plane using a UV microscope in reverse. This causes a reduction in size. By feeding the UV-emitting tube with a *brightening* pulse so that only a

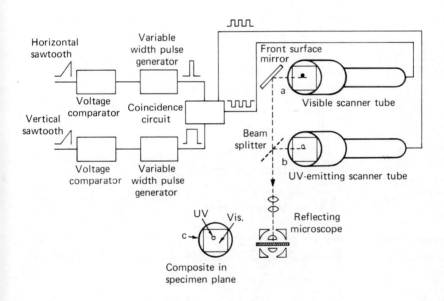

Fig. 2-5. Montgomery's flying-spot television UV microbeam. (From Montgomery and Hundley, 1961.)

small area of its surface emits UV light, the area of the emission may be only a couple of microns in diameter. The low intensity of the emission requires exposure times of several hours.

The impression given so far is that all the macrobeams that are focused to microbeams use UV light. This is not true. Laser light is routinely focused by objectives. Laser emission can occur anywhere between the far UV and the infrared portions of the electromagnetic spectrum. However, the majority of the laser microbeams have employed emissions in the visible region of the spectrum. Consequently, conventional light microscope optics can be employed readily.

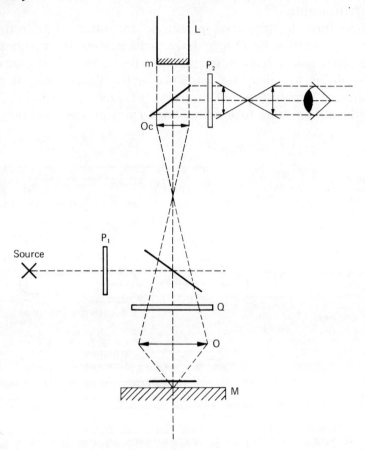

Fig. 2-6. The Bessis et al. ruby laser microbeam: *L*, laser; *M*, dielectric mirror; *M*, exit surface of laser rod; *Oc*, ocular; *O*, objective; *Q*, quarter wave plate; $P_1 P_2$, polarizers. (After Bessis, Gires, Mayer, and Nomarski, 1962.)

The first laser microbeam developed by Bessis and co-workers in 1962 (Fig. 2-6) used ruby rods 3 mm wide and 50 mm long excited by a xenon flash lamp. The laser was mounted above a microscope equipped with a vertical illuminator and a dielectric mirror on which the target specimen was placed. Alignment was achieved by observing a second image of the specimen that was projected onto the exit mirror of the laser. The brightness of the image was controlled by two polarizers and a quarter wave plate. The laser beam was passed down a 6X ocular and into a 100X objective that focused the beam to a calculated diameter of 2.5 microns.

Another ruby laser microbeam employed by Saks and Roth was reported a year later (1963) and is probably the only commercially available *packaged* microbeam system (Fig. 2-7). As in the

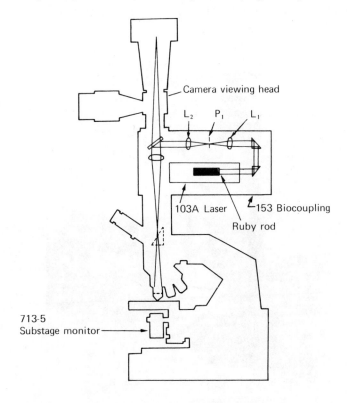

Fig. 2-7. The Hadron ruby laser microbeam system. P_1, pinhole aperture; L_1 and L_2, focusing lenses. (Compliments of Hadron Inc., Westbury, N.Y.)

Bessis-Nomarski system, the laser is mounted above the microscope and the beam is made coincident with the optical system of the microscope. The ruby crystal is 37.5 mm long and 6.25 mm wide, and produces 100 μJ in a 500 μ sec pulse. By using oculars and objectives in varying combinations the system has been used to produce focal spots varying in diameter from 62.5 to 5.3 μ. The specimen is illuminated with standard substage condensers and light, and the image is projected through the microscope system. This is the type of system that has been used most often in ruby laser microbeam studies. The system is currently marketed by Hadron (800 Shames Drive, Westbury, N.Y. 11590) and can be purchased with either a ruby rod (emission 6942 Å) or a neodymium glass rod (emission 1.06 μm). The basic laser unit can be coupled to either a standard upright microscope (Fig. 2-7) or to an inverted microscope (Fig. 2-8).

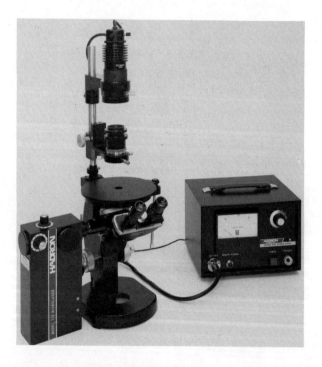

Fig. 2-8. Hadron ruby laser microbeam coupled to an inverted microscope. (Courtesy of Hadron Inc., Westbury, N.Y.)

Development of the argon ion gas laser made available wavelengths in the blue and green regions of the spectrum (4579 Å to 5145 Å). The first microbeam system with this laser was developed in the Laser Biology Department at the Pasadena Foundation for Medical Research in 1968 (Fig. 2-9). The system utilized a low power pulsed argon laser (1.5w, peak power/pulse; pulse duration, 50 μ sec) that was directed through an aperture in a rotating circular

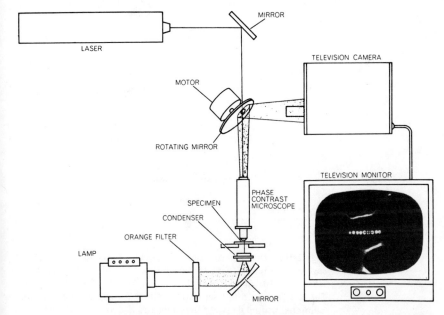

Fig. 2-9. Early argon laser microbeam. The rotating mirror sends a triggering pulse to the laser so that the laser discharges when the hole in the mirror is in a position over the microscope. (From Berns and Rounds, 1970.)

mirror (60 r/sec). An electric signal from the mirror triggered the laser to fire so that the pulse of laser light passed through the aperture and down the microscope system where it was focused by a 100X neofluar objective. The image of the target specimen was projected off the rotating mirror into a television camera, and viewed on the television screen. Irradiation was carried out by locating the cell under a cross hair on the television monitor, and then manually triggering the laser. This system permitted continual viewing of the specimen before, during, and after irradiation.

A more advanced argon laser microbeam system was developed at the University of Michigan in Ann Arbor (Fig. 2-10 and 2-11). The laser is mounted above a Zeiss photomicroscope. The beam is

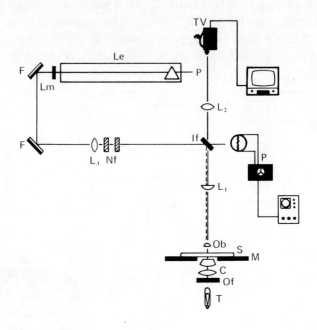

Fig. 2-10. Diagram of laser microbeam: *C,* condenser; *F,* front surfaced mirrors; *If,* interference filter; L_1, 60 mm focal length lense; L_2, 10 cm focal length lens; L_3 100 cm focal length lens; *Le,* laser cavity; *Lm,* laser output mirror; *M,* microscope stage; *Nf,* calibrated neutral density filters; *Ob,* 100 X Zeiss neufluar objective; *Of,* orange filter; *P,* photodiode, with attached photometer and oscilloscope; $\triangle$, wavelength selector prism; S, specimen chamber; *TV,* television camera and monitor; *T,* tungsten light source; ——, laser beam; –––––, sub-stage illumination. (From Berns, 1971.)

reflected at right angles by two front surface mirrors that are mounted in precision optical mounts. It is then reflected downward at a right angle by a coated interference filter that reflects 90% of the blue-green light (wavelength shorter than 520 nm), and transmits 90% of the light longer than 520 nm. The beam next passes through a 60 mm focal length lens mounted in the monocular tube of the microscope, through the vertical path of the microscope, and into an oil immersion objective (Zeiss neofluar 100X, n.a. 1.3). Light from the substage tungsten source is projected through an orange filter (transmits $10^{-4}\%$ below 520 nm) before passing through the condenser and up into the microscope system. The image of the specimen on the microscope stage is projected through the inter-

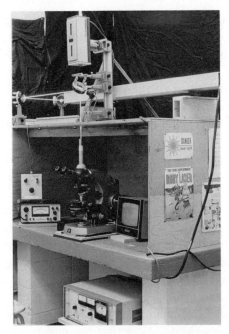

Fig. 2-11. Argon laser microbeam diagrammed in Fig. 2-10.

ference filter. A 10 cm lens is necessary to adjust the focal length of the light so that the image on the television screen is parfocal with the focal plane of the microscope. The television camera is mounted on a XY mount permitting movement in two horizontal planes.

In a normal irradiation experiment the specimen is placed on the microscope stage and viewed on the television screen. The region of the cell that is to be irradiated is moved under a cross hair on the screen, and the laser is fired. The cell may be continually observed on the television screen, before, during, and after irradiation.

The laser is a Hughes model 3030H pulsed argon laser. It can be used in single or multiwavelength operation, and in single or multimode configuration. Wavelengths are selected by rotating a prism in the rear of the laser cavity. Mode configuration can be altered by changing the output mirror. Laser output is varied by controlling the input voltage to the laser tube, or by attenuation of the beam with calibrated neutral density filters. Pulse duration can be set at either 50 or 20μ sec by varying the capacitance of the system.   Output is monitered with a calibrated S-5 vacuum photodiode mounted behind the interference filter and connected to a photometer and/or an oscilloscope. This permits recording of the actual energy of each pulse used in an experiment, as well as allowing for continual monitoring of the pulse shape.

Measurements on the system described above indicated that 12-15% of the laser energy passed into the focal spot of the oil immersion objective. By addition of a 100 cm focal length lens between the laser and the microscope the efficiency of the system was increased to 53%. The smallest effective lesion-producing spot, with or without the additional lens, was 0.5μ.

Ideally, an investigator would like one laser microbeam system that can deliver wavelengths throughout the entire visible and UV spectrum. Such a device would permit alteration of a greater variety of organelles and organisms, and it would facilitate derivation of the precise action spectrum of the particular response being studied. This capability would be invaluable for understanding the basic nature of the biological change at the molecular level.

When the organic dye laser first became available in 1971 it was felt that it would satisfy the above requirements. By simply changing the lasing medium (an organic dye) it was possible to generate an entire set of new laser wavelengths. By changing the dye and/or the concentration of the dye, it was possible to generate high energy at any wavelength in the visible region of the spectrum.

A detailed diagram of the apparatus is presented in Fig. 2-12. The five major components of the system are: (1) dye laser, (2) phase microscope, (3) recording system (television camera, monitor and videotape), (4) HeNe laser alignment system, and (5) energy recording device. Only the first two components, dye laser and microscope, are absolutely essential, though the other components provide for more efficient data collection and analysis. The dye laser either can be constructed by the investigator (Kodak Co., Rochester, New York, will provide detailed instructions on how to build a dye laser), or it can be purchased commercially. The dye is circulated from an external container (contains 300-500 ml of dye solution) through a coaxial xenon flash lamp in the laser cavity. When the lamp is fired, the dye molecules are excited and lasing action occurs between the two reflectors. The rear reflector may be either a broad-band reflecting mirror or a diffraction grating. A

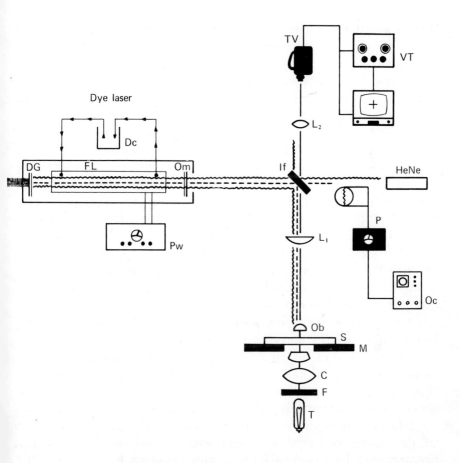

Fig. 2-12. Schematic of organic dye laser microbeam: *DC,* dye container; *DG,* diffraction grating; *FL,* flash lamp; *OM,* output mirror; *Pw,* laser power supply; *TV,* television camera; *VT,* video-tape recorder; L_2, focal length image correction lens; *If,* dichroic filter; L_1' 1X ocular; *Ob,* microscope objective (either Zeiss 40X or 100X neofluar); *S,* specimen chamber; *M,* microscope stage; *C,* microscope condenser; *F,* substage filter; *T,* tungsten light source; *HeNe,* helium neon laser alignment beam; *P,* power meter; *Oc,* oscilloscope; -----, dye laser beam; ———, substage illumination containing specimen image; ∿∿∿, alignment laser beam. (From Berns, 1972.)

broad-band reflector results in a laser emission of several hundred angstroms spread that is characteristic of the particular dye. The use of a diffraction grating permits fine wavelength tunability (±1Å). When laser wavelengths are required of a different dye or of the same dye at a different concentration, the dye container is removed and three washes of fresh solvent, usually absolute methanol, are circulated through the flashlamp. A container with the new dye solution is then placed in the system.

The microscope is a Zeiss photomicroscope with 1X ocular and either a 40X or 100X phase neofluar objective. In order to direct the laser beam down into the microscope and, at the same time, transmit the image of the biological specimen to the TV camera, a dichroic coated filter is mounted above the ocular. Three different coated filters are used, depending upon the laser wavelengths, and therefore the dye. The filters are mounted in standard optical lens holders which slip into an optical rail that is vertically mounted on a stone optical table.

The specimen image from the microscope is transmitted through the filter to a standard Sony 3200 TV camera equipped with a high sensitivity silicone diode videocon and displayed on the monitor screen. By moving the mechanical stage of the microscope the target region of the specimen can be located precisely under the cross hair on the TV screen. The TV image can be stored on ½ inch video tape by directing the TV signal through a Sony AVC 3650 video tape machine that is equipped with time-lapse capabilities of 1 frame/2 sec, and a play-back rate of 60 frames/sec. Alternately, a 16 mm or still camera may replace the TV camera.

Alignment of the laser reflectors with each other and alignment of the microscope optical system with the primary laser beam is accomplished by passing a 3 mw helium neon laser beam through the dichroic filter, down and back the cavity of the dye laser, and then down the microscope system. After alignment is complete, the HeNe beam is turned off.

The output of the dye laser is monitored with a calibrated vacuum photodiode attached to the photometer and/or an oscilloscope. Since the interference filters are only 95% reflective to the laser wavelengths, an accurate monitoring of the transmitted laser energy and, therefore, the output of the laser, is possible. Measurements in the focal plane of the 100X objective indicated that 1-10% of the total laser output was transmitted through the optical system and into the focused spot.

In the initial dye laser microbeam system four different organic dyes dissolved in absolute methanol were used: coumarin (7-diethylamino-4-methyl coumarin), fluorescein (2-dichlorofluorescein), rhodamine 6G, and acridine red. A wide variety of wavelengths from the blue to the red regions of the spectrum can be produced. Originally the only gap was between 480 and 530 nm, but now a lasing dye (B. Snavely, Kodak Co., Rochester, N.Y., personal communication) has become available for that region of the spectrum. The emission spectrum of fluorescein, rhodamine 6G,

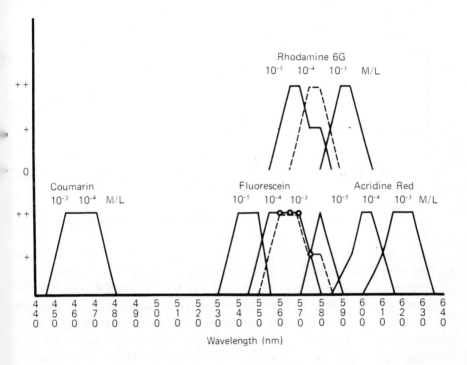

Fig. 2-13. Spectral output of various organic dyes. (From Berns, 1972.)

and acridine red can be shifted to longer wavelengths by increasing the dye concentrations (Fig. 2-13). This is an important point since inadvertent changes in concentrations due to inaccurate measuring, or aging dye solutions and solvent evaporation may result in a shift in the emission wavelengths of the laser. Energy measurements of the total laser output for the four dyes ranged from 15-200 mJ. Pulse duration for all dyes was 200 n sec therefore, at times it was possible to obtain a peak power of 1 Mw for the rhodamine dye. For a 1μ

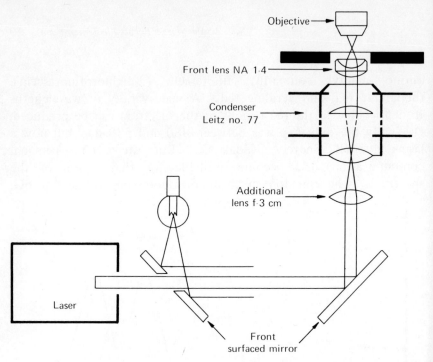

Fig. 2-14. A continuous wave helium-neon laser microbeam.
(From Lacalli and Acton, 1972.)

focused spot and a 1-10% transmission through the microscope, a total power density of 10^5 w/μ^2 is attainable.

Most recently, an economical laser microbeam has been described by Lacalli and Acton (1972). They have employed a continuous wave helium neon laser (HeNe) to irradiate cells of the plant *Microsterias*. In their system the HeNe beam is passed through a hole in a front surfaced mirror and reflected up through a short focal length lens into a condenser which focuses the laser beam to a diameter of less than 5μ (Fig. 2-14). Exposure times of a second or two are sufficient to produce damage in chloroplasts or vitally stained cells.

Other types of radiation also have been used in microbeams. A magnetic lens was used to focus electrons to a bundle which was lμm in diameter. The beam energy varied between 30-150 k.e.v. and energy rates up to 2 x 10^7 r sec^{-1} were delivered to cells (Pohlit, 1957). A problem with this system was electron scattering within the target, resulting in an effective area of damage considerably greater than lμm.

An X-ray microscope has been constructed but not applied to partial cell irradiation. It is also possible to focus other heavy

20

charged particles, such as protons, although rather large magnetic or electric fields would be needed. A much simpler and often applied method of partial cell irradiation with charged particles is to use a macrobeam with a shield.

B. MACROBEAMS USED WITH SHIELDING

A method often employed for PCI is to bombard the specimen with a macrobeam after shielding it with a strongly absorbent material, usually a metal derivative. Three possible configurations have been used (Fig. 2-15). In simplest form (a) the radiation which may be a particle, X-ray, UV, etc., is a parallel macrobeam impinging upon the unshielded portion of the specimen. A derivative of this simple approach (b) is to use a small shield (microshield) that covers

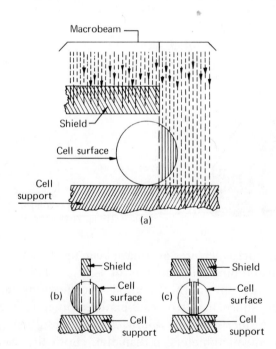

Fig. 2-15. Three methods of partial cell irradiation with shielding. Shaded areas of the cells indicate regions exposed to parallel beam. (a) unilateral shield; (b) microshield; (c) microaperature. (Modified from Zirkle, 1957.)

only a small region of the specimen. A tiny metallic bar or drop of mercury on a coverslip over the specimen has been employed. With this configuration a great portion of the cell is exposed to the irradiation while a specific region of the cell, such as a nucleus, is shielded.

The most common and versatile approach is a shield with a microaperture (c). The aperture may be produced by either placing two shields only 1 or 2 μ apart, as in (a) or by piercing a small hole through a one-piece shield. By moving either the shield or the specimen beneath the shield, the aperture can be located precisely over a desired region of the specimen.

The methods for producing microapertures attest to the ingenuity of the microbeamist. One method employed to produce a microaperture 2-5 μm in diameter from two metal plates was to scratch a small triangular groove into one of two highly polished surfaces (Fig. 2-16). When the two surfaces were pressed close together, a small aperture resulted.

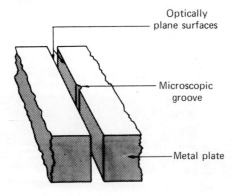

Optically
plane surfaces

Microscopic
groove

Metal plate

Fig. 2-16.   Schematic for a microaperture from two highly polished surfaces. (From Zirkle, 1957.)

Conical microapertures also have been produced by pulling lead glass capillary tubing to diameters of 1-2μm. The beam is passed down this tube. In another method, a collimating hole has been produced by squeezing two lead plates around a fiber and then extracting the fiber from the *sandwich*. The resulting aperture may be only several micrometers in diameter. Both the lead-glass capillary microaperture and the *lead-sandwich* aperture have been used to collimate X-ray beams.

Another method is to take copper electron microscope grids (produced by electroforming) with holes 6μm in diameter and by deposition evaporation carefully fill in the holes with tin until diameters as small as lμm are obtained.

Numerous methods for puncturing small holes through shields have been developed. Conical microapertures l-30μm in diameter have been made in aluminum or lead sheets (20-50μm thick) by puncturing with finely pulled glass microneedles. Electric sparks have been used to puncture small holes (l0-20μm) in thin sheets of mica, glass, and metal.

Despite all the ingenious methods for producing microapertures, it is practically impossible to produce apertures of uniform size and shape. Another major problem in using microapertures is the geometry of the irradiation pattern. When an accelerated beam of parallel particles is used as in Fig. 2-15, the radiation passes cleanly through the aperture and exposes only the desired region of the specimen. However, when the radiation source is an extended source (non-parallel particles) coming through the aperture at an

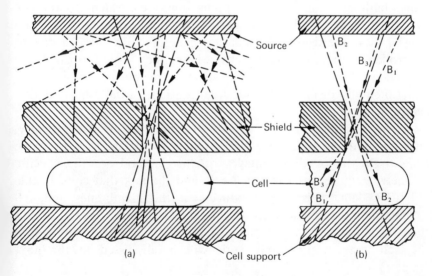

Fig. 2-17. Scheme for partial cell irradiation with randomly directed particles and shielding. (a) Randomly directed emission of particles from an extended source and the spatial relations among source, cell, and microaperture in shield. (b) Illustrates certain contralocalizing factors. B_1 particle that is not stopped but only slowed down by the shield and proceeds into the cell outside the aperture field. B_2, B_3' particles are like B_1' except that they deviate more widely because of nuclear scattering. (Modified from Zirkle, 1957.)

angle, the particles can be reflected off the sides of the aperture, or just pass through at an angle, thus extending the area of irradiation considerably (Fig. 2-17). This latter problem makes accurate determination of the radiation-damage area almost impossible and often unpredictable.

C. MACROBEAM ATTENUATED BY CELL

An infrequently used but simple method of partial cell irradiation is to expose a cell to a macrobeam of particles that impinge upon it from a wide variety of angles (Fig. 2-18a-c). The success of this technique relies upon the penetrance of the radiation and the location of the target structure. By choosing a radiation source where all of the particles have the same degree of penetrance (e.g., monoenergetic particles from polonium 210), and utilizing either bathing media with a stopping power equivalent to the cell cytoplasm (Fig. 2-18b), or a partial shield to attenuate the beam (Fig. 2-18c), the degree of radiation penetrance in the cell can be controlled. By irradiating cells where the desired organelle is located, either naturally in a specific region of the cell, or experimentally displaced to a desired location, partial cell irradiation can be achieved.

The advantages of this scheme of irradiation are its simplicity, and the ability to irradiate a large number of specimens at the same time. The major disadvantages of this method are the specific geometry required of the target cell, and the fact that considerable radiation damage is also produced in the membranes and cell cytoplasm above and adjacent to the target structure. In addition to these disadvantages, the pattern of radiation bombardment within the cell varies considerably (this will be discussed in the next chapter).

If a radiation source is used that is exponentially absorbed, such as UV, rather than having a finite range of penetrance as in the previous example using alpha particles, partial cell irradiation also can be attained. In this case the internal bombardment pattern, radiation distribution within the cell, is determined by the dimensions and composition of the cell. These factors determine the linear absorption coefficient within the target. In a situation wh

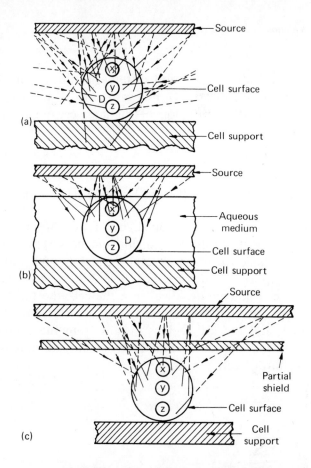

Fig. 2-18. Partial cell irradiation with extended source of randomly directed monoenergetic particles. (a) cell in medium of low stopping power (e.g., air); (b) in aqueous medium; (c) in air, with partially absorbing shield. *x, y, z*, various positions that might be occupied by some cell organelle or other target structures. (Modified from Zirkle, 1957.)

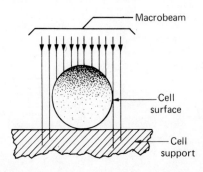

Fig. 2-19. Differential cell irradiation with an exponentially absorbed macrobeam. Shading indicates quantitative degree of radiation absorption. (From Zirkle, 1957.)

25

there is a quantitative absorption differential throughout the cell, from top to bottom, (Fig. 2-19), PCI is attained. Such a system has been employed in the irradiation of the large eggs of *Habrobracon* (see Chap. 6).

D.   MICROSOURCE EXTERNAL TO CELL

A unique approach developed by T. R. Munro (1957) at Cambridge University utilizes a small source, several microns in diameter, of radioactive material (alpha emitter, such as polonium 210) on the tip of a microneedle (Fig. 2-20). By placing the radiation source a known distance from the cell, the finite range of the particles insures that the regions of the cell closer to the source are irradiated, and the cell regions distant from the source escape irradiation. This approach has been employed extensively in studying the mitotic cell.

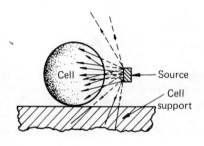

Fig. 2-20. Arrangement for partial cell irradiation with a microsource of radiation of definite range. (From Zirkle, 1957.)

Radiation Classification and Sources
THREE

A. RADIATION CLASSIFICATION

Radiation can be divided into two major groups: particulate and electromagnetic. Radiations from both groups are used in microbeams.

Particulate Radiation

The particulate radiations consist of atomic particles that are classified as *light* or *heavy*. The light particles are either positively or negatively charged. The negatively charged electron is the particle emitted from matter when sufficient energy becomes available to overcome the attraction of the positively charged nucleus. This can occur by heating or by stimulating with some other radiation. A cathode is the negative, electron emitting terminal of an electric circuit in a vacuum vessel, commonly called a cathode ray tube. Electrons are also referred to as negative beta particles. The other major type of light particle is the positively charged *positron*. Since the charge is equal to the negative charge of the electron, this particle has been referred to as the positive electron, or positive beta particle.

27

The heavy particulate radiations are positively charged atoms that have been stripped of some or all of their electrons. The proton is the bare hydrogen nucleus (at. wt. 1.008) with a positive charge equal to one electron. The a particle is the bare nucleus of helium (at. wt. 4.003) with a positive charge equal to two electrons. In addition to the charged particles, there are also neutral particles, or neutrons, which occur in the nuclei of all atoms except for the common isotope of hydrogen.

The effects of particulate radiation are produced by collision of the particles with matter. The *potency* of these radiations, therefore, is expressed in terms of the kinetic energy of the particles. The unit of energy measurement is the electron volt (ev), which is equal to 1.6×10^{-19} J or 1.6×10^{-12} erg.

The kinetic energy of the particle is dependent upon both its mass and its velocity. This relationship is contained in the expression, $v = 1.4 \times 10^6 \sqrt{E/M}$ cm/sec, where v = velocity, M = mass, and E = kinetic energy in electron volts. This expression indicates the importance of considering the number of particles and their velocity when determining the potency of a particulate radiation source. It also suggests why many partial cell irradiation devices employ beams of accelerated particles, since more energy is contained in an accelerated particle.

Electromagnetic Radiation

Electromagnetic radiation generally is classified by either wavelength (λ) in angstroms, or by frequency (v) in cycles/sec. Both of these are related in the expression $\lambda = c/v$, where c equals the velocity of propagation of the radiation ($c = 3 \times 10^{10}$ cm/sec). The potency of electromagnetic radiation is expressed in terms of energy imparted to electrons that are emitted from matter when struck by the radiation. Each ejected electron receives a definite amount of energy that is specific to the frequency of the impinging radiation. The energy that electromagnetic radiation imparts to an electron equals the product Lv, where L is Planck's constant (6.62×10^{-27} erg/sec, or 4.14×10^{15} ev/sec) and v is the frequency (cycles/sec) of the radiation. For example, UV light with a wavelength of 2500Å (2.4×10^{-5} cm) has a frequency (v) equal to c/λ or $3 \times 10^{10}/2.5 \times$

10^{-5} cm which equals 1.25×10^{15} cycles/sec. The energy, Lv, that radiation of the wavelength imparts to each emitted electron equals 4.14×10^{-15} (1.25×10^{15}) or 4.96 ev. The amount of energy (Lv) that is imparted to an electron when radiation of a specific wavelength strikes matter, is called a *photon*, or a *radiation quantum*. It should be pointed out that radiation of low frequency does not contain enough energy to result in ejection of electrons from matter. One must also remember that the ability of electrons to be emitted from matter is going to depend upon the binding forces within the matter itself. The lowest frequency of radiation capable of ejecting electrons from a material is called the *photo-electric threshold* of that material. The frequencies, wavelengths,

Type	Wavelength (λ) nm	Frequency (υ) cycles/sec	Electron volts ev
Gamma rays	0.0001	10^{23}	10^8
X-rays	0.01	10^{21}	10^5
UV	200	1.5×10^{15}	6.25
Near UV	350	0.0857×10^{15}	3.57
Visible			
blue	450	0.0667×10^{15}	2.78
green	500	0.0600×10^{15}	2.50
yellow	600	0.0500×10^{15}	2.08
red	700	0.0429×10^{15}	1.79
Infrared	750	0.0400×10^{15}	1.67
	1150	0.0261×10^{15}	1.09

Fig. 3-1. Comparison of electromagnetic radiation.

and photon energies of the different types of electromagnetic radiation are summarized in Fig. 3-1. It should be noted that as wavelength increases, frequency decreases, and the photon energy decreases. One must be careful in comparing the energy of particulate radiation with electromagnetic radiation, since the former varies considerably depending upon the velocity of the particles.

B.   RADIATION SOURCES

Particulate

Particulate radiations are produced by both natural processes (radioactive decay) and manipulative design. Negatively charged electrons most routinely are generated in a cathode ray vacuum tube (Fig. 3-2). A typical rate of production is 10^{+9} e/sec. Protons are obtained by stripping electrons from gas atoms at low pressure. The stripping results from the strong atomic collisions accompanying an electric discharge.

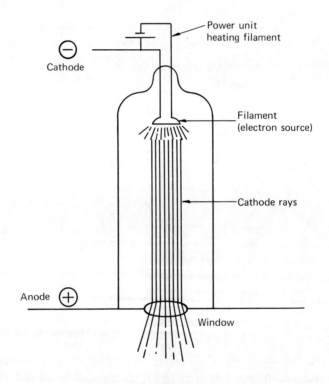

Fig. 3-2.   Schematic of cathode ray tube. (Modified from *Principles of Radiological Physics* by U. Fano. 1954, McGraw-Hill. Used with permission of McGraw-Hill Book Company.)

All charged particles can be formed into a beam by attracting them towards a conductor of the opposite charge. They can then be focused into beams of small diameter by subjecting them to appropriate electric or magnetic forces. As discussed in the previous chapter, an electron beam has been produced with a diameter as small as $1\,\mu$, and an energy range of 30-150 Kev.

Since the kinetic energy of a particle increases with its acceleration, numerous methods for accelerating particles have been devised. The normal method of acceleration in a vacuum cathode ray tube is to subject the charged particles to an electric force field between two conductors with a large potential difference. Another method employs the Van de Graaf generator which uses a mechanically driven insulator belt to pick up electric charges at one point and deposit them onto a high voltage conductor (Fig. 3-3). A rather unique heavy particle microbeam employed a 2 Mev Van de Graaf generator to accelerate a beam of deuterons (bare nucleus of

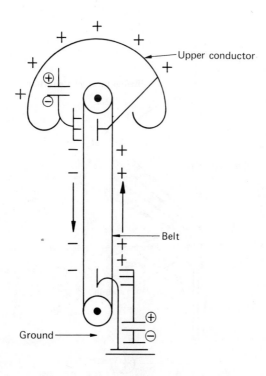

Fig. 3-3. Van de Graaf generator. Moving belt picks up positive charges, deposits them on upper conductor, and carries negative charges down to ground. Potential differences as great as 12,000,000 v are possible. (Modified from *Principles of Radiological Physics* by U. Fano. 1954, McGraw-Hill. Used with permission of McGraw-Hill Book Company.)

the heavy isotope of hydrogen, at. wt. 2.015, positive charge of one electron) into a chamber of He^3. The result was a nuclear disintegration between the deuteron and H^3, producing 15-16 Mev protons which were then collimated and passed into the target.

Neutrons become available as byproducts of nuclear collisions and remain available for only a short time because they are readily captured by any nucleus. A standard method of producing neutrons is to bombard atoms with a beam of accelerated deuterons, using a Van de Graaf generator, thus causing the release of the loosely bound deuteron neutrons. Because of their short lifetime, neutron beams are not practical for partial cell irradiation.

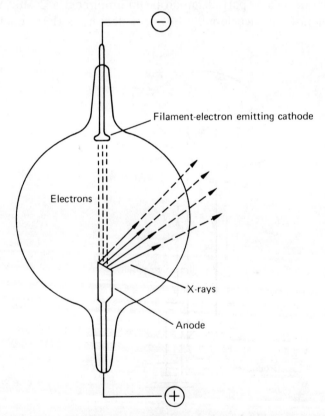

Fig. 3-4. Schematic of X-ray tube using electrons to bombard anode materials such as tungsten, aluminum, molybdenum. (Modified from *Principles of Radiological Physics* by U. Fano. 1954, McGraw-Hill. Used with permission of McGraw-Hill Book Company.)

Electromagnetic

There are three conventional sources for electromagnetic radiation in the UV, visible, and infrared regions of the spectrum: (1) thermal radiators, (2) electrical discharge in a gaseous environment, (3) fluorescent tubes. Since most partial cell irradiation devices have employed gaseous discharge lamps, only superficial coverage will be given to thermal radiators and fluorescent lamps. In addition to the conventional sources, *Light Amplification by Stimulated Emission of Radiation* (LASER) is also possible in the UV, visible, and infrared regions of the spectrum. The short wavelength X-rays and gamma rays are produced either as emissions from atomic nuclei, or by stimulating appropriate materials with electrons in a conventional X-ray tube (Fig. 3-4).

Thermal Radiators

Thermal radiators are incandescent lamps. In these lamps electrical power is used to heat a tungsten filament, which emits a spectrum of temperature-dependent wavelengths (Fig. 3-5). Incan-

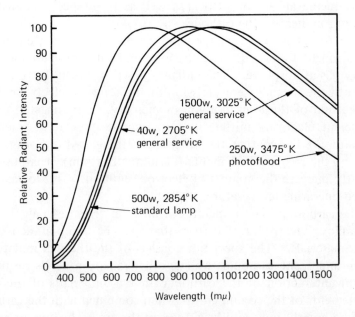

Fig. 3-5. Spectral emission of various types of tungsten filament lamps with filament temperatures indicated. (From Forsythe and Adams, *J. Optical Soc. of America* **35**: 108-113, 1945.)

descent lamps are poor sources of radiation below 3000Å and in fact, only 10% of their emissions are in the visible region of the spectrum. Most are in the infrared, as is true of all thermal sources of radiation. Though incandescent lamps generally are not used as the source for radiation in a microbeam system, they are frequently used for illumination of the specimen. The large amount of infrared energy produced by those lamps necessitates the use of a long wavelength, heat-absorbing filter.

Gaseous Discharge Lamps

There are many types of gaseous discharge lamps. By varying such parameters as the type of gas, gas pressure, operating current, and cathode materials, it is possible to produce a wide variety of emissions.

Emission is produced as a result of electrons being accelerated across a potential gradient in a gaseous atmosphere. Characteristic wavelengths are produced from the electrons striking the atoms of gas (producing an arc or spark), as well as from spectral emission resulting from heating the cathode.

The tungsten arc discharge lamp employs a tungsten electrode that is brought to incandescence by ion bombardment in an argon or mercury gas atmosphere. The radiant energy from this lamp comes from both the incandescent emission of the tungsten electrodes and the spectrum of the ionized argon or mercury. For tungsten to be an efficient electron emitter, it must operate close to its melting temperature. By coating the tungsten cathode with thorium, or oxides of the rare earth elements (barium, strontium), it is possible to greatly increase the number of electrons emitted. This lowers the required operating temperature.

The carbon arc lamp operates at medium pressure in air (1 atm) and employs electrodes of carbon that may be compounded with various materials. The spectrum consists of the thermal radiations from incandescent carbon as well as other radiation bands resulting from the interaction of the carbon with the air. Salts of the rare earth elements of the cerium group, when combined with the carbon, result in a greatly increased brightness of the arc. The low intensity carbon arc employs potassium combined with carbon, and emits strongly in the blue and near UV, operating at a current of 10-30 amp. The high intensity carbon arc combines carbon with the cerium salts. Most of the radiant energy comes from the highly incandescent vapor cloud of cerium at an operating current of 125-150 amp. The carbon flame arc lamp actually relies upon the burning of the cathode. By combining with the carbon the elements that emit specific wavelengths when ignited, specific spectral emissions can be obtained: calcium-yellow, strontium-red, metals such as iron, nickel, cobalt, aluminum-ultraviolet.

The zirconium arc operates on a dc current between a zirconium metal cathode and a refractory metal anode. The arc is initiated by a high voltage discharge in an argon gas atmosphere. The spectral emission is that of a thermal radiator with the spectrum of argon and zirconium superimposed (near UV, 3000 Å to infrared).

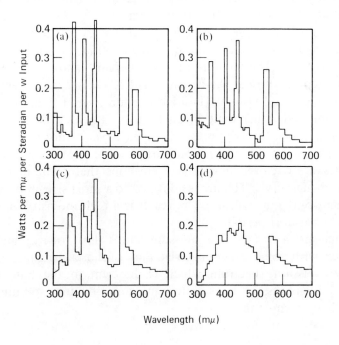

Fig. 3-6. The effect of mercury vapor pressure on the spectral energy distribution of high pressure mercury arcs. (a) 54 atm, (b) 102 atm, (c) 197 atm, (d) 319 atm. (From Withrow and Withrow, 1956.)

An arc discharge in a hydrogen gas atmosphere produces emission from the UV through the visible. These emissions can be produced at low gas pressures and high voltage, or at high gas pressures and low voltage.   A typical hydrogen lamp operates at 0.2 mm of hydrogen at 50-20 amp.   The radiation is rather diffuse and of low intensity.

The xenon arc lamp is rich in blue and UV emissions, operates at 20 atm, and utilizes tungsten electrodes.   A uniform and broad spectrum is attained with the incandescent tungsten spectrum superimposed over the xenon spectrum (2000-2300 Å).

The mercury arc discharge lamp is probably the most widely employed.   Because the distribution of energy into the various excited mercury atoms depends upon the vapor pressure, mercury lamps of low, medium, and high pressures have been developed.   At low pressure (0.008-0.1 mm Hg) the distance between the mercury atoms is so large that the probability is great that the energy will be radiated at 2437 Å − the transition from the lowest excited state to the ground state.   Even though the efficiency of conversion of electrical energy to radiant energy is high (60%) the overall output intensity is low.   The medium pressure mercury lamps operate at an increased temperature at 0.5-10 atm.   The result is a greater emission in the near UV and visible. The high pressure mercury lamps (30 to several hundred atm) emit strongly over a wide spectral range depending upon the Hg pressure (Fig. 3-6).   Another type of mercury lamp is the *short* concentrated arc.   Massive tungsten electrodes are enclosed in a quartz envelope that contains a small amount of mercury.   The arc is confined to a short stream of 10 mm length and 5-10 mm width.   The result is a high intensity emission confined to a short period.

Amalgam arcs are high pressure lamps (100 atm) of mercury combined with some other material that is chosen to augment the mercury emission spectrum which is deficient in the long wavelengths.   Elements, such as cadmium, zinc, thalium, sodium, and cesium, have been used.

Fluorescent Lamps

In fluorescent lamps the UV emission of a low pressure mercury discharge (2537 Å) is absorbed by a phosphor that coats the inside of the fluorescent tube and is re-emitted at longer wavelengths.   The

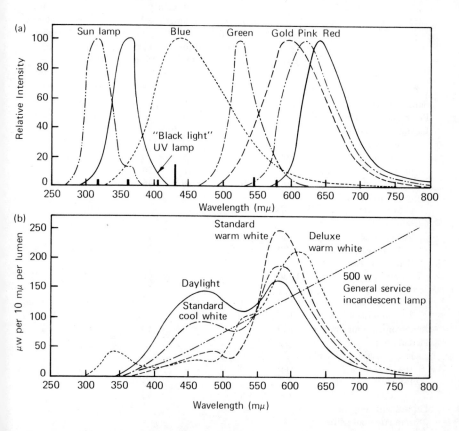

Fig. 3-7. Spectral emission of flourescent lamps; (a) emission of various phosphors; (b) various lamps available. The mercury lines are not indicated. (From Withrow and Withrow, 1956. Data courtesy of General Electric Company.)

spectral emission of the fluorescent lamp depends upon the particular phosphor used (Fig. 3-7a). For example, the common white fluorescent lamps combine a variety of blue-green and red-emitting phosphors (Fig. 3-7b). The entire emission of a fluorescent lamp consists of the phosphor emission spectrum superimposed over the weak spectrum of the mercury discharge.

Because of the general low intensity of the fluorescent lamps, they are not practical for partial cell irradiation. Figure 3-8 compares some of the general characteristics of the various types of electromagnetic radiation sources.

Source	Power, w	Brightness cmm^{-2}	Luminous efficiency, lumens w^{-1} Absolute	Luminous efficiency, lumens w^{-1} Electrical	Color temp., °K	Wave length of maximum intensity, mμ
Sun, air mass = 2	–	1650	100	–	5000	500
Black body, 3000°K	–	28	19	–	3000	1040
Incandescent lamp						
general service	40	4.5	17	11	2760	1020
general service	500	9	23	20	2960	1000
projection	500	16	30	26	3190	900
photoflood	1000	30	40	36	3360	850
Tungsten arc, G.E.						
photomicrographic	330	48	46	–	3700	700
Carbon arc						
low intensity, 30 amp	1650	150	–	13	3600	1000
high intensity, 180 amp	13300	950	–	25	6000	600
flame, rare-earth, 40 amp	1500	–	–	60	4700	400
Zirconium arc						
0.085 mm diam. source	2	96	–	–	–	1000
1.50 mm diam. source	100	52	–	–	–	1000
Xenon arc, 3.5 mm gap	150	170	–	18	4000	1050
Mercury arc						
low pressure, germicidal	30	–	–	–	–	254
H-4, 8 atm	100	18	–	27	–	365
H-6, 110 atm, water-cooled	1000	300	–	65	–	436
Fluorescent lamp						
standard cool white, 60 in.	85	0.008	110	55	4500	590
green, 48 in.	40	0.01	150	75	–	530
Sodium lamp	180	0.06	–	53	–	589

Fig. 3-8. Comparison of various electromagnetic radiation sources. (From *Generation, Control, and Measurement of Visible and Near-Visible Radiant Energy* by R. B. Withrow and A. P. Withrow. 1954, McGraw-Hill. Used with permission of McGraw-Hill Book Company.)

Lasers

The revolutionary technique of stimulated emission of radiation was theoretically conceived by the Russians Basov, Prokhorov, and an American Charles Townes, and resulted in all three receiving the Nobel prize in Physics. The initial predictions actually were made for microwaves, thus the term *M*icrowave *A*mplification by *S*timulated *E*mission of *R*adiation (MASER). In 1958 Charles Townes of Columbia University and Arthur Schwalow of Stanford University predicted that stimulated emission could occur in the visible region of the spectrum, thus the term *optical maser* was born. This term has given way to *L*aser — *L*ight *A*mplification by *S*timulated *E*mission of *R*adiation.

The first laser action was produced by Theodore Maiman, originally with Hughes Aircraft Company, in 1960. This first system was exceptionally simple, employing a ruby crystal with partially silvered ends, a flash lamp, and a power supply (see Fig. 3-9).

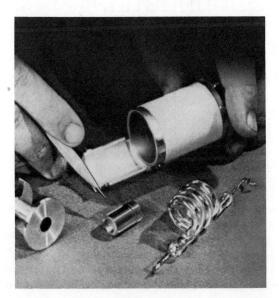

Fig. 3-9. Actual components of first ruby laser. Pencil indicates ruby crystal; flashlamp to right of crystal. (Compliments of Korad Division of Union Carbide, Inc.)

Perhaps the simplest way to illustrate what makes laser light so different than light emitted from a conventional lamp is to describe a typical laser instrument. The basic features of a ruby laser are diagrammed in Fig. 3-10. The key to attaining laser action in the ruby crystal is to excite a large number of the chromium atoms to higher energy levels. This is accomplished by discharging the flash lamp that surrounds the rod. Since chromium absorbs a broad band of green, yellow, and UV light, many of the photons emitted when the flash lamp is discharged are absorbed by the chromium atoms. The result is an *inversion* of the population of chromium atoms from the ground (lowest) energy state, to an excited energy state (Fig. 3-11

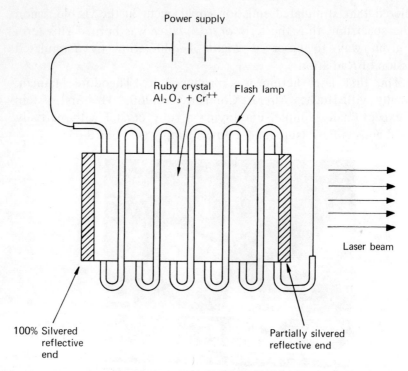

Fig. 3-10.   Schematic of a simple ruby laser.

a, b).   From here two steps are required to carry a chromium atom back down to the ground state (Fig. 3-ll c, d).   In the first step the excited atom loses some of its energy to the crystal lattice and drops down to an intermediate energy level, called the metastable state, where it remains for several milliseconds before spontaneously dropping back down to the ground state.   When it makes the transition from the metastable state to the ground state, a photon with a wavelength of 6943Å is emitted as fluorescence.   This would be the normal course of events if only a few of the chromium atoms had been excited to the higher energy levels.   However, in the situation where a large number of chromium atoms have been excited (resulting in a population inversion) by a very intense burst of light, the probability becomes great that a fluorescent photon will strike a chromium atom that is still in its metastable state.   When this occurs, the metastable chromium atom is stimulated to *drop* down to the ground state, resulting in the emission of a second

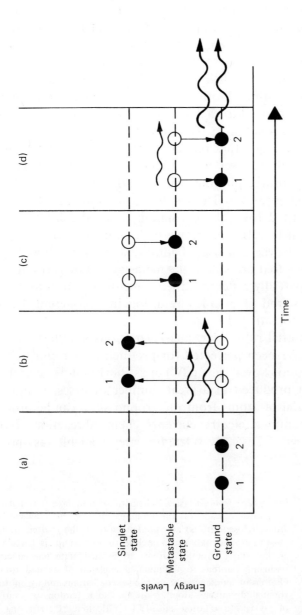

Fig. 3-11. Energy transitions characteristic of chromium atoms in stimulated emissions. (a) two ground state chromium atoms 1, 2; (b) excitation to singlet excited state by flashlamp photons; (c) transition to metastable state; (d) atom 1 spontaneously drops to ground state emitting a photon that stimulates atom 2 to drop to ground state. Both photons from atom 1 and atom 2 are in phase.

41

photon.   The fluorescent photon that stimulated the metastable atom is re-emitted completely in phase with the second stimulated photon.   The result is two photons with the same wavelength traveling parallel and in perfect phase (Fig. 3-11d).   This process is called stimulated emission and explains three of the four fundamental properties of laser light: (1) temporal coherence (one or a few discrete wavelengths of light), (2) spatial coherence (photons traveling in phase with each other), (3) very low divergence (photons traveling virtually parallel to each other).   The fourth feature of laser light, the enormous intensity (photon density) of the beam, is a direct result of placing either reflective mirrors or reflective coatings at the two ends of the laser rod.   If the two reflective surfaces are parallel to each other, photons will be reflected back and forth between them (Fig. 3-12).   The result is a photon cascade.   As the photons are emitted from the chromium perpendicular to the two mirrors, they will be reflected between the mirrors along the axis of the rod stimulating the emission of additional photons along the way by striking atoms that are still in the metastable state.   If one of the mirrors is only partially reflective, then some of the photons will pass out of the ruby rod (Fig. 3-12) resulting in a coherent beam of intense, monochromatic light.

Since the initial ruby laser, many variations have been developed.   Methods have been perfected for producing laser emission that are either continuous over time (CW) or as short as 10^{-12} sec.   Lasing action has been produced in a great number of gases, liquids, and solids.   Excitation of atoms from the ground state can be produced by flash lamp pumping, electric currents, chemical reactions, or even by another laser.   Laser wavelengths are available as primary

Fig. 3-12.   Opposite page. Schematic of ruby laser action illustrating photon cascade. (a) Unstimulated ruby crystal; a few chromium atoms (closed circles) are spontaneously in the excited state, but most in ground state (open circles). (b) Flashlamp discharge excites ground state chromium atoms to excited state (closed circles); flashlamp photons. (c) Stimulated emission of excited atoms by photons of spontaneously emitting excited atoms and by photons of stimulated emitting atoms. Note that some photons pass out sides of crystal. (d) Photon cascade by reflection off reflecting ends; stimulated emission continues. (e) Further photon cascade and passage out partially reflective end.

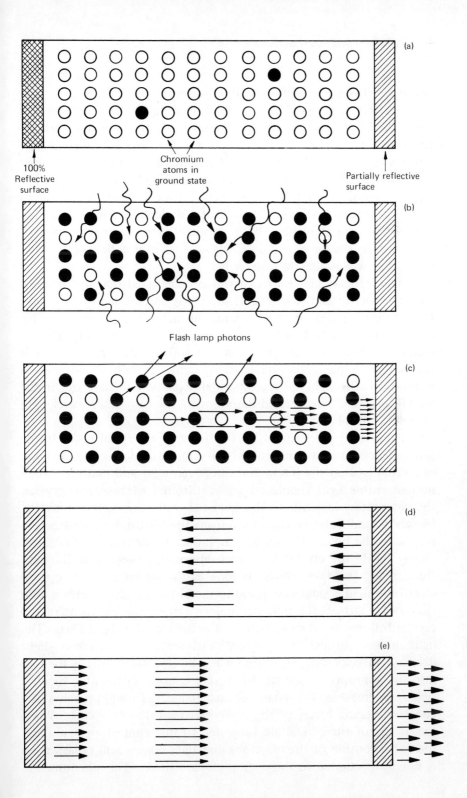

(a)

100%
Reflective
surface

Chromium
atoms in
ground state

Partially reflective
surface

(b)

Flash lamp photons

(c)

(d)

(e)

emissions from the infrared to the near UV (3200 Å). In combination with nonlinear optical techniques it is possible to generate the second or third harmonics of many of the primary emission wavelengths, thus extending the available laser wavelengths down into the far UV. This possibility was alluded to as far back as 1845 and 1875 in the early experiments of Michael Faraday and the Scottish physicist John Kerr. However, it was entirely impossible to test their theories because the intensity of electromagnetic radiation required was unattainable. With laser light it has become possible to demonstrate nonlinear optic effects and to utilize these effects to generate new wavelengths of laser light.

The Faraday and Kerr experiments suggested that by virtue of its own electric and magnetic field, highly intense light can change the refractive index of the medium through which it passes. This can, in turn, affect the propagation of the light through the medium. When a laser wavelength of sufficiently high intensity is focused into an appropriate transparent crystalline material, a portion of the light is re-emitted at exactly one half the initial wavelength (e.g., the second harmonic of the ruby laser wavelength of 6943 Å is 3471.5 Å).

How can light focused into a transparent crystal result in the generation of a new wavelength? To understand this phenomenon one must look at the interactions between light and matter. When normal visible light (nonlaser) passes through a transparent crystal, the alternating electric field associated with the light causes the loosely bound outer (valence) electrons to redistribute themsleves in step with the field. The result is an induced polarization of negative charge inside the crystal in a weak alternating current at the light frequency. In other words, as long as the strength of the optical electronic field is small compared to the binding electric fields within the crystal lattice, the polarization pattern of the electrons of the crystal follows the electric field of the light wave (Fig. 3-13a). The light passes through the crystal unchanged except for a slight decrease in velocity. However, when intense laser light is focused into an appropriate crystal, the result is quite different. Electric fields as great as 10^7 v/cm are attainable. This approaches the cohesive electric forces of the crystal lattice (10^8-10^{10} v/cm). When the atoms of the crystal are subjected to this kind of electric field, the redistribution of the electrons and their consequent polarization is not proportioned to the optical electric field. This phenomenon

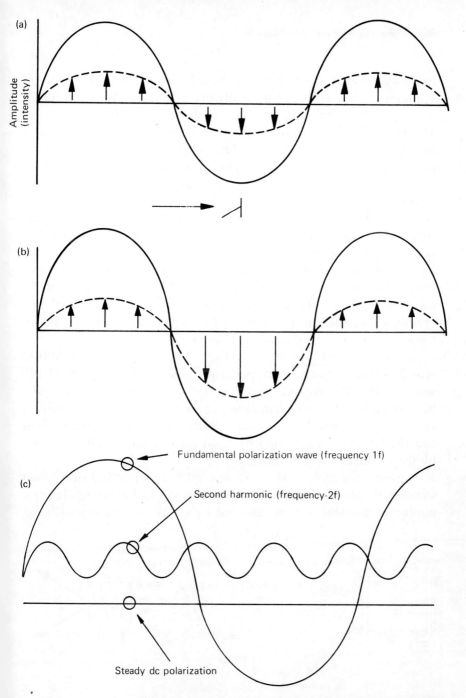

Fig. 3-13. Generation of nonlinear polarization wave. (a) light wave of moderate intensity (solid line), and linear oscillating polarization wave of the crystal electronic charge (dashed line); (b) same wavelength of light, of high intensity with nonlinear polarization wave; (c) three components of nonlinear polarization wave.

can only occur if the *appropriate* crystal does not have any center of symmetry.   About 10% of naturally occurring crystals have this type of asymmetry, and they are referred to as *piezoelectric* materials. The resulting wave produced in these crystals is called a *nonlinear polarization wave* (Fig. 3-13b).   The nonlinear polarization wave has been shown to have three components, two of which are emitted separately as waves.   One wave has the frequency of the initial laser wavelength, and the other wave has twice the frequency of the initial laser light.   Both waves are emitted as parallel, coherent beams.   The third component is a steady polarization, or direct current, that does not result in light emission (Fig. 3-13c).   It is possible, therefore, by passing an intense laser beam into a *piezoelectric* crystal, to generate laser light with twice the frequency (half the wavelength) of the incident radiation.   This approach already has been utilized in a partial cell irradiation device in the Paris laboratory of Marcel Bessis, Christian Salet, and Giuliana Moreno.   They have focused a neodymium doped glass laser (principal output wavelength 10,600 Å) into a piezoelectric crystal of potasium dihydrogen phosphate (KDP) and obtained the second harmonic, 5300 Å green light.   They also have obtained the third harmonic, 2650 Å by focusing the 5300 Å light into a second KDP crystal.   Both the green second harmonic and the UV third harmonic have been used in partial cell irradiation by these researchers (Fig. 3-14).

Another technique that can be used to generate additional laser wavelengths is stimulated Raman scattering.   In Raman scattering a photon of incident light is absorbed by a molecule and re-emitted at

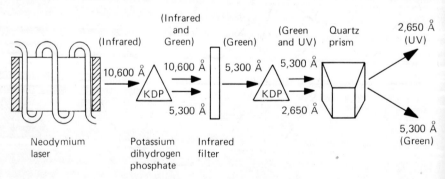

Fig. 3-14.   Second and third harmonic generation by passing a laser beam through the piezoelectric crystal potassium dihydrogen phosphate (KDP).

a lower frequency (longer wavelength) after losing a small amount of energy in the form of energy of vibration or rotation. The Raman effect was discovered with conventional light in 1927 by the Indian physicist C. V. Raman. Though very distinct wavelengths could be produced by Raman scattering, the intensities were quite low. However, by coupling a Raman scattering material with a laser, the Raman wavelengths are generated as an intense coherent beam. The process is very similar to the method of laser action. If enough Raman scattered photons are produced and reflected between mirrors, they can stimulate further Raman emission from the molecules as they resonate between the mirrors. A large number of Raman lines become available by varying the Raman scattering material, and the stimulating laser wavelength. By incorporating nonlinear crystals, such as KDP, into the path of the beam it is possible to produce an interaction between the Raman line and the original laser wavelength. These interactions result in the intense monochromatic beams *beating* together to produce new frequencies that correspond to the sum and difference of the interacting beams. Thus, the number of available coherent wavelengths become enormous.

It is clear that the number of laser wavelengths available for partial cell irradiation is limitless. However, the application of frequency generation by Raman scattering, harmonic generation, or a combination of both, is often not easy for the biologist because of

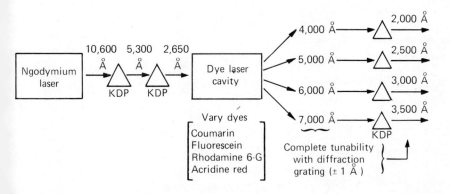

Fig. 3-15. Theoretical wavelength versatility by combining a dye laser with harmonic generation of both dye laser wavelengths and the stimulating laser wavelength.

expense and physical sophistication. More recently, the commercial availability of the organic dye laser provides a source of laser radiation with a limitless number of wavelengths. For example, the use of just three dyes will provide a broad band of wavelengths from the blue to the infrared region of the spectrum. Fine tuning is attained by placing a diffraction grating as the rear mirror of the laser cavity. Combination with frequency shifting crystal should provide limitless wavelengths in the UV (Fig. 3-15).

Radiation, Matter, and Dosimetry
FOUR

All radiations contain energy. It is the dissipation of this energy that is responsible for changing matter. Since the basic units of matter are atoms and molecules, it is the interaction of radiation with these units that is critical. The action of radiations on atoms and molecules generally is analyzed in terms of specific *elementary processes*.

A. ELEMENTARY PROCESSES

Compton Scattering

Electromagnetic radiation causes charged particles (such as electrons) to oscillate according to the variation of the electric force characteristic of the radiation. The oscillations produced in the charged particles result in the propagation of a variable current which is radiated out in many directions. When an electron is struck by incident radiation of sufficient energy, such as high energy X-ray photons, the electron is ejected from the atom and the incident photon is scattered (re-emitted) in a different direction.

Pair Production

When electromagnetic radiation of high energy (1 Mev X-ray photon) is propagated near a charged particle, the result is an outright ejection of an electron and a positron. The energy of the absorbed X-ray photon equals the combined energy of the two ejected particles. A small amount of energy is lost to the charged particles in the actual collision.

Elastic Collision of Charged Particles (Rutherford Scattering)

When an electrically charged particulate radiation approaches another charged particle, the incident charged particle is deflected or attracted off its original path. This tends to scatter the particles of the incident beam and is termed an *elastic collision.*

Inelastic Collisions

These involve the excitation or ionization of an atom or molecule and result from the direct interaction of the incident particle with the atomic electrons. It is the inelastic collisions that account for the major actions of all the high-energy radiations.

Magnetic Forces

Electrically charged particles in motion generate magnetic forces and are themselves subject to magnetic forces produced by other currents. The magnetic effects increase in proportion to the speed of the particles.

B.   ACTIONS OF SPECIFIC RADIATIONS

Electromagnetic

As mentioned previously, electromagnetic radiation may act on atomic systems by precipitating internal oscillating currents. The induction of currents may result in a loss of energy from the incident radiation and its redistribution in the target system in a process called *absorption*, or by a re-radiation in a different direction as *scattering.* An example of the latter process is *Compton scattering*, and, if the incident radiation is high energy (i.e., a high energy X-ray photon), it may result in the ejection of an electron from the target, as well as the re-emission of the stimulating photon. It is possible for a low-energy X-ray photon to be scattered, but not result in the ejection of an electron. This occurs when the recoil of the electron is not great enough to eject it from the atom or molecule. Such a process would not result in a change to the target system. Other than the ejection of electrons via Compton scattering, the primary action of electromagnetic radiation is through *absorption*.

The major effect of an absorbed photon is on the orbital electrons. If an absorbed photon is of sufficiently high energy, such as an X-ray photon, the outer electron may be ejected entirely from the atom in an ionization process. However, if the photon is of lower energy, UV or visible wavelengths, the result may be excitation of an orbital electron from its regular orbit to a higher energy level orbit, thus raising the atom from its ground state to the singlet excited state (Fig. 4-la). Once in the excited singlet state, the atom will rid itself of the excess energy by one of several processes. The most common and direct process is the re-emission of a photon as fluorescence. The fluorescent photon is emitted when the excited electron drops back down to its original orbit, resulting in the atom dropping from the excited singlet state down to its ground state. The energy of the fluorescent photon is lower (therefore, wavelength is longer) than the stimulating photon. This is due to the fact that when an atom is excited initially, it goes to a higher vibrational level within the singlet state. However, before the atom drops down to the ground state, it drops down to the lowest possible vibrational level of the singlet state. The excess vibrational energy of the excited singlet state is dissipated as heat by collisions with other atoms or water, or whatever makes up the solvent.

Another process that results in the emission of a photon is phosphorescence. Certain molecules with highly conjugated structures can radiate photons from an electronic state (the triplet state) between the ground state and the fluorescent singlet state. The lifetime of the triplet state is unusually long (10^{-3} sec), and therefore, increases the probability that the excited reactive molecule will participate in a chemical reaction. The lifetime of the excited singlet state is about 10^{-6} sec.

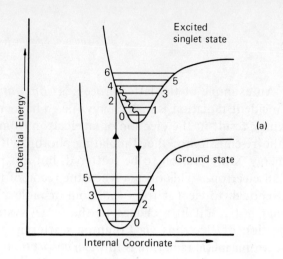

(a)

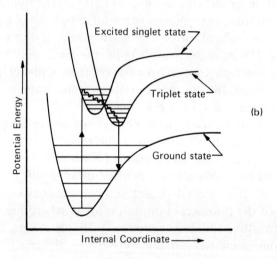

(b)

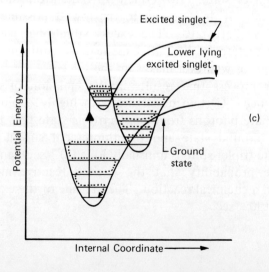

(c)

For phosphorescence to occur, an atom is first excited to a high vibrational level of the singlet state. It drops down to a lower vibrational level of the singlet state (losing vibrational energy on the way) and then crosses over to a high vibrational level of the triplet state. The atom passes through the vibrational levels of the triplet state before dropping to the ground state and emitting a phosphorescent photon (Fig. 4-1b). It is possible also for an excited molecule to dissipate energy by a nonradiative process (light is not re-emitted). In this process the excitation energy is dissipated by conversion entirely to vibrational energy. For this to occur there must be either a second, lower excited singlet state, or a triplet state that overlaps with the ground state. Energy, then, is dissipated merely by progressive vibrational loss until the ground state is attained (Fig. 4-1c). The result of *vibrational de-excitation* is a greatly heated molecule, since vibrational energy loss is a thermal process. The probability of the heated molecule participating in a chemical reaction is greatly increased.

Vibrational de-excitation is also the standard mechanism of energy dissipation when a molecule is stimulated by a low energy photon (visible or infrared photon). The molecule can be stimulated to higher vibrational levels only within the ground state. Energy dissipation will be entirely by vibrational heating. Considerable thermal damage could be produced in a target molecule by continual irradiation with low energy photons.

Fig. 4-1. Opposite page. Energy diagram of excited atomic states. (a) A diagrammatic representation of photon absorption and fluorescence. A representative internal coordinate for a diatomic molecule might be internuclear distance. A potential well might then schematically represent the possible internuclear separations for a given electronic state of the molecule as shown in this figure. The minimum energy value or bottom of the well indicates the equilibrium internuclear separation. Numbered bands within these potential wells represent the different vibrational states of the molecule. (Rotational energy states could have been indicated in this figure as a fine structure of bands between the various vibrational levels.) At higher vibrational energy states the variation in internuclear separation increases until sufficient energy is attained to overcome the attraction between the nuclei; this results in the breaking apart of the molecule. Note that the minimum internuclear distance approaches a limiting value as the vibrational energy increases. Vibrational deexcitation in the excited electronic state is indicated by the wavy line. (b) Potential energy diagramatic representation for triplet state excitation and phosphorescence. (c) The nonradiative process of vibrational deexcitation and internal conversion. (From Smith and Hanawalt, 1969.)

Charged Particles

The interaction of charged particles with atoms or molecules generally involves three types of collisions: elastic, inelastic, or radiative. As discussed earlier in this chapter, these reactions may result in considerable modification to the target atoms. Thus, the biological effects of charged particle radiation are derived primarily from disturbance of the atomic structure of matter. The most common effect is ionization resulting from inelastic collisions. The electric force exerted by an incident charged particle upon each atomic electron increases as the particle approaches and decreases as the particle moves away. If the incident particle is traveling with sufficient speed and passes close enough to the atomic electron, the disturbance resembles a sudden blow. This is called a *fast* collision and it may result in the ejection of an electron (ionization) if the charged particle passes close to the target atom (*knock on* collision), or it may result in a more indirect effect (*glancing blow*) if it passes far away from an atom. In the latter situation the electrons are excited to one of their higher energy levels and then drop back down to the ground state, and in the former case they may be excited enough to be ejected from the atom.

Slow collisions occur when the electric force increases and decreases slowly. The atomic electrons have time to drop back down to the ground state. This type of interaction occurs when the incident particles travel at relatively slow velocities.

A frequent action of high-speed charged particles on matter is a succession of *fast* inelastic collisions. Often large numbers of inelastic collisions take place along the path of a charged particle, resulting in a progressive degradation of the energy of the charged particle. These interactions may be regarded as *glancing collisions* and result in the transfer of energy at the level that precipitates a chemical change, rather than the outright ionization that results from a direct *knock on* collision.

C.   GEOMETRY AND DOSIMETRY

In the preceding section we discussed the various types of radiations and the ways they affect biological atoms and molecules. The discussion to this point has dealt with the general effects of various radiations in a nonmicrobeam situation. The concentration of these various radiations to a small spot 1μ or so in diameter generally does not affect their basic mechanisms of action. The only difference is that the volume of biological material affected is smaller.

However, it is the irradiation of such a small volume of biological material that causes considerable problems in determining the radiation dose. The biological potency of a particular radiation depends in part upon the concentration of the radiation. Determination and prediction of the radiation dose is often crucial to the successful microbeam experiment. Two approaches may be taken: (1) theoretical calculation of energy dose, or (2) actual measurement. Both approaches require precise knowledge of the geometry of the irradiation area. This information is often difficult to obtain. In the case of *particle* microbeams the high energy particles frequently are focused by finely drawn capillary tubes or passed through small apertures in an absorbing material. In both cases, particles may be scattered off the inner surfaces of the delimiting material or the specimen itself, resulting in a *damage area* several microns greater than the aperture or capillary bore.

Problems of dosimetry with X-ray microbeams also exist, but they are not as great as in the particulate systems. Because of the low energy flux in a small beam of X-rays, a two stage approach is taken. First, the potency of the broad beam is measured with a standard dosimeter. Second, a fine grain autoradiographic film emulsion is exposed to the broad beam in one portion, and another portion of the emulsion is exposed to the microbeam. Identical photographic development of the two films provides an indirect calibration of the microbeam. Another approach to X-ray microbeam dosimetry has been the use of the LD_{90} of *Drosphila* eggs (Seidel and Bucholtz 1960) as a biological assay.

UV microbeam dosimetry usually involves the use of a calibrated detector whose readout is corrected according to the absorption properties of the target. Detectors such as thermocouples, bolometers, thermistors, photoelectric receivers, vacuum photocells, photoresistors, and photographic emulsions have been employed. These devices may be used to record the radiation dose prior to entry into the focusing system. From the calculations of spot diameter and attenuation by the optical system, energy density at the focused spot can be estimated. However, unless the absorption characteristics of the target region are precisely known, the real potency of the radiation cannot be determined. Since the absorption spectra of various regions of the cell vary considerably, depending upon the relative concentration and occurrence of various UV-absorbing molecules (proteins, nucleic acids), the real potency of the radiation may be impossible to determine.

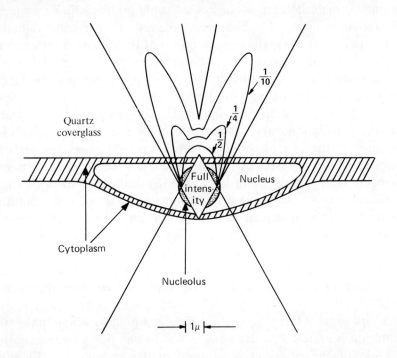

Fig. 4-2. Idealized profile of a cell being bombarded by an UV microbeam which is focused on its nucleolus. The cell geometry is estimated from measurements on fixed material. Above and below the bicone of full intensity are shown isointensity contours of diminishing intensity. (From Perry, 1961.)

In addition, the geometry of the focused beam must be taken into consideration. For example, an UV beam focused by a reflecting objective appears as a hollow cone of intensity converging to a spot which is determined by the pinhole diameter and the objective magnification. The angle of convergence is determined by the numerical aperture of the objective and the refractive index of the specimen. The geometry of such a focused beam is illustrated in Fig. 4-2. In this particular situation it has been estimated that 70%

of the radiation dose is absorbed by the nucleolus, the primary target, and 30% by cytoplasm and nucleus (Perry, 1961). It is obvious that the best results are obtained with flattened, thin cells.

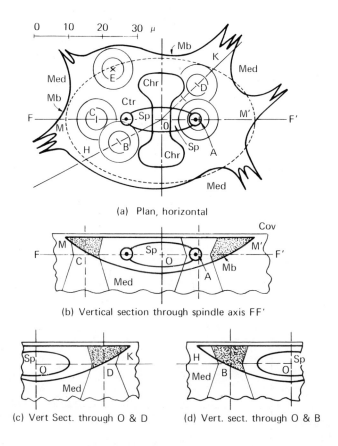

(a) Plan, horizontal

(b) Vertical section through spindle axis FF'

(c) Vert Sect. through O & D   (d) Vert. sect. through O & B

Fig. 4-3.   Schematic relations of a metaphase *Triturus* cell and five regions traversed by conical microbeams. Cell parts shown by heavy lines. *Sp*, spindle; *Ctr*, centriole, surrounded by centrosome; *Chr*, region occupied by chromosomes; *Mb*, cell membrane; *Med*, culture medium; *Cov*, bottom surface of coverslip; *O*, intersection of spindle axes; *A, B, C, D, E*, centers of focal spots of microbeams focused on plane through spindle with axis *FF´* and parallel to coverslip. (a) Dashed curve is an ellipse with axis of 50 and 75 μ, taken as formal horizontal cell area and used in constructing (b), (c), (d). Each pair of concentric circles represents the cross-sectional areas of a microbeam as it enters the cell (15μ diameter) and in the focal plane (8μ diameter). (b, c, d) Stippling denotes cell volumes traversed by variously located microbeams. (From Zirkle, 1970.)

An excellent example of a microbeam experiment employing flat cells and containing the geometrical information necessary for precise interpretation of the experimental results was Zirkle's (1970) study on UV irradiation of the mitotic spindle in rat heart cells. Figure 4-3 illustrates the type of geometrical definition presented in this study. Whereas this kind of determination may not always be possible, it does add a certain degree of confidence to the experimental results.

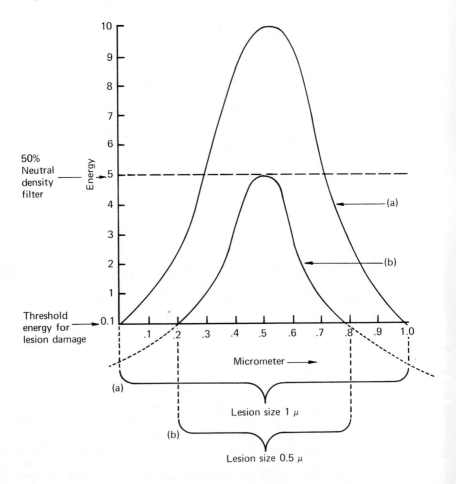

Fig. 4-4. Reduction of lesion size by calibrated attenuation of focused laser spot with a gaussian energy distribution: (a) no attenuation, lesion diameter, 1μ, (b) 50% attenuation, 0.5μ lesion diameter.

Visible light (laser) microbeams are monitored much in the same way as UV microbeams. However, because the radiation is from the visible portion of the spectrum, the absorption coefficients of various regions of the cell are more easily determined. Only a few visible-absorbing molecules routinely occur in the cell and they are located in only a few specific organelles (chlorophyll in chloroplasts, cytochromes in mitochondria). In addition, organelle-specific vital dyes may be used to sensitize particular regions to radiation of

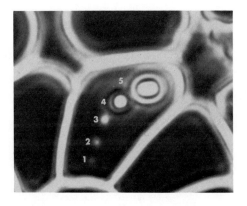

Figure 4-5. Actual lesions in human red blood cell illustrating lesion size reduction when the laser beam is attenuated. The total laser output energy was the same for each irradiation. However, a different calibrated neutral density filter was used to attenuate the beam for each lesion: lesion 1, 0.48% filter (transmission); 2, 0.62%; 3, 0.79%; 4, 1.0%; 5, 1.15%. Lesion diameter: 1, 0.25μm; 2, 0.40μm; 3, 0.50μm; 4, 0.90μm; 5, 1.4μm.

specific wavelengths. It, therefore, becomes possible to define the absorption characteristic of cells. Because the energy density across a focused laser spot follows a normal gaussian distribution, precise uniformity and predictability of dose can be expected. This factor also permits the production of a lesion considerably smaller than the diameter of the focused spot. Consider a 1μ spot with a gaussian energy distribution across it (Fig. 4-4a). Assume that the threshold for biological damage is 0.1 unit on the energy scale. If neutral density filters are placed in front of the beam the total energy would be reduced such that a portion of the gaussian drops below the threshold for biological damage.   In Fig. 4-4b a 50% (transmission) neutral density filter is placed in front of the beam, only $\frac{1}{2} \mu$ falls above the 1 energy level (threshold for biological damage).   The rest of the focused spot (dashed line) falls below the threshold.   It has been possible to obtain lesions down to 0.25 μ by using this approach (Fig. 4-5).   It should be emphasized that this capability exists only when a gaussian energy distribution is obtained.   Such a feature is characteristic of *single mode* lasers.

The Cell

FIVE

The largest number of microbeam experiments has involved microirradiation of cells. These have been either cells in culture or unicellular organisms. A detailed discussion of all of these would be time-consuming and tedious. Consequently, I will discuss a group of investigations selected to illustrate the variety of problems studied and how they have contributed to a better understanding of the cell.

The target region of the cell is either the nucleus or the cytoplasm. Specific structures or regions within these areas may be irradiated, or the entire nucleus or cytoplasm may be irradiated.

A. WHOLE NUCLEAR OR
WHOLE CYTOPLASMIC IRRADIATION

The earliest studies involving irradiation of either the entire nucleus or cytoplasm were performed on various animal and plant embryos and will be the topic of discussion in the next chapter. A favorite target in nonembryonic systems has been the Protozoan *Amoeba*. Several investigators have irradiated whole, enucleated, and nucleated fragments of amoeba with ionizing (X-ray) and UV microbeams. The general goals of these investigations have been to study the role of the nucleus and cytoplasm in cell function, specifically in cell division, as well as the interactions between the nucleus and cytoplasm.   In their 1969 paper, Jagger, Prescott, and Gaulden discussed the results of their UV studies in relation to earlier studies by other investigators.   Even though the earlier studies indicated that UV or ionizing effects on cell survival and mitotic

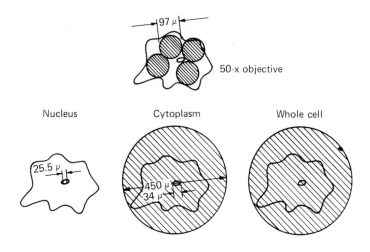

Fig. 5-1.   Diagrams of irradiation fields used with the 50X objective (top) and 20X objective (bottom). Cross-hatched areas indicate the beams used. In the lower cytoplasmic irradiation scheme, the cell was entirely irradiated except for a *shadow* spot 34μ in diameter. (From Jagger et al., 1969.)

delay were about the same for nuclear or cytoplasmic irradiation, the techniques employed made the results suspect.   For example, in the studies where nuclei from irradiated (X-ray or UV) cells were transferred into unirradiated cytoplasm and unirradiated nuclei were transferred into irradiated cytoplasm, the time between irradiation and actual transfer of the nucleus varied from 15 min. to 8 hr.   Thus, it would have been impossible to detect any nuclear-cytoplasmic interactions occurring within that time period.   In other studies where nucleate, enucleate, and whole amoebas were irradiated in the unflattened *free* state, it is pointed out that the rounded-up state of the organism at the time of irradiation undoubtedly permitted considerable shielding of the nucleus by the cytoplasm.   In the

Jagger, Prescott, and Gaulden studies the amoebas were artificially flattened to a 20 μ thickness. This permitted irradiation of either the whole nucleus or cytoplasm without the necessity for microsurgery and transplantation, a procedure that could greatly alter the metabolism and biochemistry of the cell (Fig. 5-1). The results indicated that in flattened amoebas cell death resulted equally from irradiation of nucleus or cytoplasm. However, division delay was obtained much more easily with cytoplasmic irradiation (Fig. 5-2).

Object tive	Beam diameter (μ)	Cell part irradiated	Dose for division delay (b)	Reciprocal (100 ×)	Dose for killing(c)	Reciprocal (100 ×)
50 ×	97	Cytoplasm ($\sim$50%)	15000 (43000)		85000	
20 ×	25.5	Nucleus ($\sim$70%)	4200	0.24	3900	0.26
	450	Cytoplasm ($\sim$98%)	690	1.45	4200	0.24
	450	Whole cell (N + 100% C)	470	2.12	1350	0.74
		Interaction factor (d): 2.12/1.69 = 1.25			0.74/0.50 = 1.48	

Fig 5-2. UV doses to various cell parts required for equal effects. (a) Incident energy per unit area in units of erg/mm^2; (b) Doubling of the time required to complete the second division; (c) 50% lysis at 21 days; (d) Reciprocal of dose for whole cell divided by sum of reciprocals of doses for nucleus and cytoplasm. (From Jagger et al., 1969.)

Using these observations, the authors performed a series of experiments designed to test whether or not the cytoplasm of free unflattened amoebas protected the nucleus from ultraviolet microirradation. They found that division delay, which is caused primarily by cytoplasmic damage, is produced as easily in free cells as flattened cells (Fig. 5-3), but killing, which is caused equally by cytoplasm or nuclear irradiation, is difficult to produce in free cells but easy to produce in flattened cells (Fig. 5-4). These results do suggest that the cytoplasm shields the nucleus from UV irradiation. The same authors also looked at photoreactivation (reversal of observed effect when exposed to longer wavelength light, 330-440 nm) of both nuclear and cytoplasmic irradiation. Surprisingly, they found that nuclear irradiation was only weakly photoreactivable, but

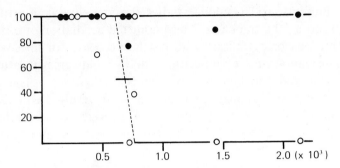

Abscissa: UV dose (erg mm⁻²); ordinate: % survival.

Fig. 5-3.   Survival following germicidal lamp irradiation of whole *A. proteus* as free cells (•) or as cells flattened to a 20μ thickness (o). The broken line is drawn to fit the open symbols and show a LD50 of 670 erg/mm^2. Abscissa: UV dose (erg mm^2); ordinate: % survival. (From Jagger et al., 1969.)

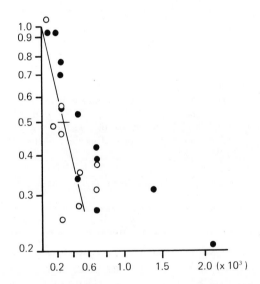

Fig. 5-4.   Abscissa: UV dose (erg mm^{-2}); ordinate: division-delay factor. Division delay factor as a function of UV dose, following germicidal lamp irradiation of whole *A. proteus* as free (unflattened) cells (•) or as cells flattened to a 20 μ thickness (o). Data are from four experiments, each involving both free and flattened cells. The straight line is drawn to fit the open symbols, and shows a dose of 300 erg/mm^2 for a division-delay factor of 0.50. (From Jagger et al., 1969.)

that both the killing and division delay response of cytoplasmic irradiation was highly photoreactivable. The authors apparently demonstrated that cytoplasmic damage that has wide implications for the entire cell is photoreactivable-a phenomenon that had been indirectly suggested by many authors, but never convincingly demonstrated.

Amoebas were used also in UV microirradiation experiments by Skreb (1960). Oxygen consumption of whole cells, separated nuclei, and enucleated cells was compared to oxygen consumption in UV-microirradiated nuclei and enucleated cytoplasm. Irradiation of nuclei had no effect on oxygen consumption, but irradiation of the enucleated cells reduced oxygen consumption. In a later study, the same cell components were UV-microirradiated, and RNA and protein synthesis studied. Protein synthesis was less affected with nuclear irradiation than with the cytoplasm irradiation. The microbeam studies in amoeba suggested specific biochemical functions to both nucleus and cytoplasm and permitted inquiry into the nature of nuclear-cytoplasmic interactions that might very well be common to all cells.

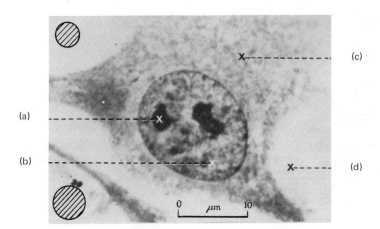

Fig. 5-5. Photograph showing a typical cell and the sites at which the particle and UV microbeam would be directed. (a) irradiation of the nucleolus; (b) irradiation of an area of nuclear sap which contains no nucleolar material; (c) irradiation in the cytoplasm; (d) irradiation of the medium surrounding the cell. (From Dendy and Smith in *Proc. Roy. Soc.,* 1964.)

The preceding studies were some of the first attempts to employ microbeam irradiation in the elucidation of the biochemical interaction between nucleus and cytoplasm. A more detailed series of experiments designed to examine the biochemical relationships between the nucleus and cytoplasm were those of Dendy (1962), using a UV microbeam, Smith (1961, 1964), using an alpha particle device, and Dendy and Smith (1964) using both methods. Several different regions of embryonic mouse fibroblasts and mouse L-cells in culture were irradiated with both types of microbeams: (1) nucleolus, (2) nuclear sap, (3) cytoplasm, and (4) medium surrounding cells (Fig. 5-5). The cells were incubated in H^3-thymidine at various times after irradiation and then analyzed autoradiographically. The irradiation of all four regions resulted in some inhibition of DNA synthesis. However, since the reduction in DNA synthesis was the same for nucleolar and nuclear irradiation, it was concluded that the nucleolus does not play a significant role in DNA synthesis.

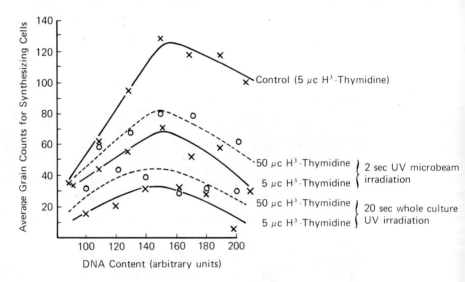

Fig. 5-6.   Variation of rate of DNA synthesis plotted against DNA content for UV microirradiated cells. (From Smith, 1964.)

Inhibition of DNA synthesis by cytoplasmic and culture medium irradiation was attributed to the *probable* production of a nuclear toxin by the UV radiation. However, studies with the a particles alone did not result in an inhibition of DNA synthesis following cytoplasmic irradiation. In a subsequent study Dendy and Cleaver (1964) investigated the effect of UV microbeam irradiation on the rate of DNA synthesis at different stages during the DNA synthetic phase. They measured spectrophotometrically the relative amount of DNA from a large number of DNA-feulgen-stained cells and correlated this with incorporation of H^3-thymidine. These values were compared with similar values obtained from cells that were whole-cell or nuclear-irradiated with a UV microbeam (Figs. 5-6 and 5-7). Differential sensitivity to microbeam irradiation during the cell cycle was demonstrated.

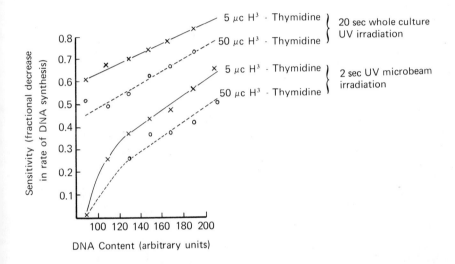

Fig. 5-7. Variation in sensitivity of DNA synthesis to UV microirradiation during position in *S* phase. (From Smith, 1964.)

But perhaps some of the most intriguing studies in which the microbeam has been employed to study the nucleus and cytoplasm are those of Henry Harris and his colleagues at Oxford. In their early studies it had been possible to genetically reactiviate a hen erythrocyte nucleus that had been incorporated into the cytoplasm of mouse cells by the cell fusion technique. The erythrocyte nucleus, which normally does not transcribe its genome, became functional after a period of time in the mouse cell. Furthermore, some of the RNA produced by the hen nucleus moved to the cytoplasm of the mouse cell and presumably directed the synthesis of hen-specific proteins.   These hen proteins were assembled with the protein-synthesizing machinery of the mouse cell!

In the preceding studies, it was observed repeatedly that before the RNA moved out of the hen nucleus a nucleolus was produced in the hen nucleus, which is normally nucleolus-deficient.   This

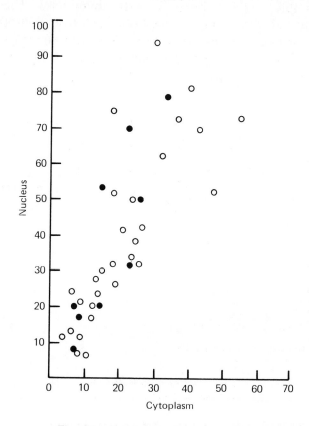

Fig. 5-8.   Relationship between nuclear grain counts and cytoplasmic grain counts in heterokaryons and in single mouse cells. The heterokaryons contained one mouse nucleus and up to four chick erythrocyte nuclei that had been reactivated by UV light both in the heterokaryons and in the single mouse cells. The cells were exposed to a radioactive RNA precursor for 6 hr. The ratio of nuclear to cytoplasmic RNA labeling in the heterokaryons in which the mouse nucleus had been inactivated was no different from that in the single mouse cells in which the nucleus had been inactivated. The reactivated erythrocyte nuclei, although they synthesize large amounts of RNA, do not make any detectable contribution to cytoplasmic RNA labeling at this stage. O, heterokaryons; •, single mouse cells. (From Harris, 1970.)

observation prompted the theory that *some nucleolar activity was essential for the transport of this RNA to the cytoplasm of the cell.* A UV microbeam was used to investigate this hypothesis.

One of the first experiments was to UV-inactivate the mouse nucleus of the fused hybrid cell, that is, a cell containing nuclei of both mouse and hen. The levels of cytoplasmic and nuclear RNA were measured by exposing the irradiated cells to a radioactive RNA precursor for up to 6 hr. and then making an autoradiograph. The cells chosen always had one mouse nucleus (the irradiated one) and one or several hen nuclei. Measurements of RNA were made at different times after cell fusion: (1) before any of the erythrocyte

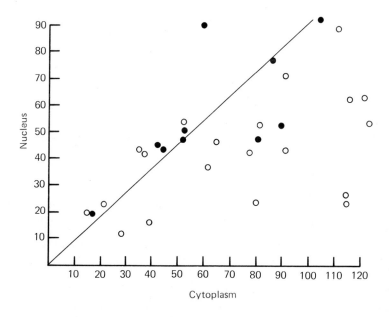

Fig. 5-9. An intermediate stage when some of the erythrocyte nuclei had begun to develop nucleoli. It will be seen that the cytoplasmic labeling is now greater in some of the heterokaryons than in the single mouse cells. The reactivated erythrocyte nuclei are beginning to contribute to cytoplasmic RNA labeling. O, heterokaryons; •, single mouse cells. (From Harris, 1970.)

nuclei formed nucleoli (Fig. 5-8); (2) after some nucleoli had been formed (Fig. 5-9); (3) after several days when all the hen nuclei had well-formed nucleoli (Fig. 5-10). The control experiment was to

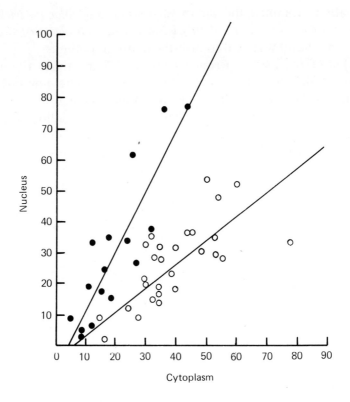

Fig. 5-10. A stage when virtually all the erythrocyte nuclei had developed well-defined nucleole. The cytoplasmic labeling in the heterokaryons is now decisively greater than in the single mouse cells. The erythrocyte nuclei are making a substantial contribution to cytoplasmic RNA labeling. O, heterokaryons; •, single mouse cells. (From Harris, 1970.)

irradiate the nuclei of mononucleate mouse cells not fused with hen cells and detect the amount of cytoplasmic label.   This was necessary since a low level of cytoplasmic RNA label still persists after nuclear inactivation.   During the time when the hen cell nuclei had not yet developed nucleoli (condition 1 above), the level of cytoplasmic labeling was low (the same as in the control), even though the several hen nuclei had synthesized a relatively large amount of RNA as suggested by extremely heavy nuclear label (Fig. 5-11).   When fused cells with hen nuclei in the early stages of nucleolar formation were incubated in labeled precursor (condition 2) the amount of cytoplasmic label was significantly greater than the

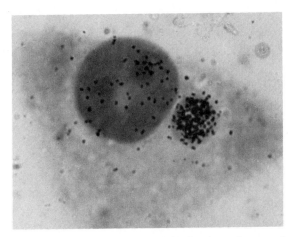

Fig. 5-11. Autoradiograph of a heterokaryon where the mouse nucleus has been inactivated by UV light and contains only moderate levels of label. The erythrocyte nucleus, which has not yet developed a nucleolus, is heavily labeled, but the cytoplasm contains very little radioactivity. (From Harris, 1970.)

control. Several days after fusion, when the hen nuclei had well developed nucleoli (condition 3) the amount of cytoplasmic label was greater than both the control and the previous cells with early stage nucleoli (Fig. 5-12). From these experiments it is concluded that the nucleolus is indeed necessary for the movement of

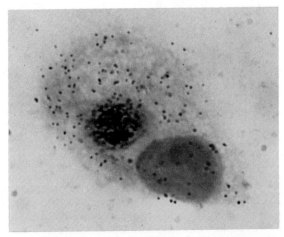

Fig. 5-12. Autoradiograph of a heterokaryon after the erythrocyte nucleus had developed a nucleolus. The cytoplasm of the cell is now labeled. (From Harris, 1970.)

polydisperse RNA or *messenger* RNA from the nucleus to the cytoplasm. It should be pointed out that it was important to inactivate the mouse nucleus in order to eliminate the possibility that it either contributed detectible labeled RNA to the cytoplasm or in some other way participated in the movement of the RNA from the hen nucleus to the cytoplasm.

In a later series of experiments, Harris irradiated mononucleate (nonfused) HeLa cells to test the theory in another system. Nuclear

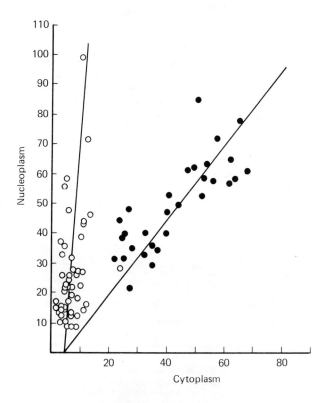

Fig. 5-13. Comparison of the number of grains over the cytoplasm with the number of grains over the nonnucleolar parts of the nucleus (nucleoplasm) in unirradiated HeLa cells. (•) and in HeLa cells in which the nucleolus only was irradiated (o). The cells were exposed to a radioactive RNA precursor for 6 hr. Irradiation of the nucleolus alone abolishes cytoplasmic RNA labeling, despite the fact that RNA synthesis continues in the rest of the nucleus. (From Harris, 1970.)

and cytoplasmic labeling were assayed autoradiographically in the following: (1) unirradiated cells, (2) cells with a single nucleolus that had been UV inactivated (Fig. 5-13), (3) cells in which the nonnucleolar portion of the nucleus was irradiated, and (4) cells in which the entire nucleus was inactivated. The results demonstrated that the amount of cytoplasmic RNA labeling was reduced by about 90% when either the nucleolus or the whole nucleus was inactivated, but hardly reduced at all when the nonnucleolar nucleoplasm was irradiated. These results agree with the previous studies on the fused cells, and, furthermore, indicate that the effect was not due to a nonspecific consequence of radiation since nucleoplasm irradiation did not affect cytoplasmic RNA labeling. Harris feels that both the mouse-hen and the HeLa studies indicate that the nucleolus is involved in the transfer to the cytoplasm of both nucleolar RNA (ribosomal) and the information-containing RNA made elsewhere in the nucleus. Two major alternatives to this hypothesis have been proposed: (1) that the observed results really reflect the fact that the ribosomes, which are made by the nucleolus, are necessary in the cytoplasm for the translation of messenger RNA; (2) the production of a nucleolus and concomittant translation of messenger RNA to protein, are unconnected events that occur at the same time. In the first alternative, the messenger RNA of the donor nucleus wait in the cytoplasm until the nucleolus of the donor is formed, and therefore donor ribosomes become available for protein synthesis. This theory has been discredited by numerous cell culture and cell fractionation studies that demonstrate a nonspecific specificity of ribosomes for protein synthesis (e.g., mouse ribosomes can be used to translate hen messenger RNA to protein). The second alternative was approached in a recent study by Deak, Sidebottom, and Harris (1972). In this study the nucleoli of the reactivated hen nucleus were irradiated after they were well formed, and after the enzyme isosinic acid pyrophosphorylase was being produced by the heterokaryon. Following nucleolar inactivation, the enzyme system deteriorated substantially. Similarly, studies on the production of the chick-specific surface antigen also demonstrated a drop in antigen production following UV microbeam inactivation of the erythrocyte nucleoli. Both of these investigations strongly indicate that the nucleolus does participate in transport of messenger RNA from nucleus to cytoplasm.

B. PARTIAL NUCLEAR
IRRADIATION–NUCLEOLUS

The nucleolus is an ideal target for partial cell irradiation. Because of its well-defined morphology, spherical nature, and large size (one to several microns in diameter) it is easily located and irradiated. As discussed in the previous section, Harris has UV-microirradiated nucleoli in mouse-hen hybrid cells and in HeLa cells, in order to study the mechanisms of RNA movement from nucleus to cytoplasm. Similarly, Dendy and Smith irradiated the nucleolus with UV and alpha particles to study its role in DNA synthesis.

The nucleolus is an organelle whose functions are not well understood. Indeed, only in the past 10-12 years have the roles of the nucleolus in the cell been intensively studied. Many of these investigations have employed microbeam irradiation. From these studies several functional roles of the nucleolus have been hypothesized. The most commonly agreed upon function is that it is the major site of ribosomal RNA synthesis. In the early 1960's Perry, Hell, and Errera (1961) demonstrated that UV microirradiation of the nucleolus in HeLa tissue culture cells resulted in a marked reduction of nuclear RNA synthesis and a consequent $\frac{2}{3}$ reduction in cytoplasmic RNA (Fig. 5-14a). In a subsequent study, Errera demonstrated that there was also a reduction of cytoplasmic protein synthesis following nucleolar microirradiation. All of these studies employed radioisotope incubation and autoradiography following microbeam irradiation. (See Fig. 5-14b for experimental outline.) Similar investigations using argon laser microbeam irradiation of nucleoli have demonstrated a marked reduction of RNA synthesis in the CMP (adenocarcinoma) tissue culture cell (Fig. 5-15) (Berns et al., 1969). The sum total of these investigations resulted in the conclusion that the nucleolus produces a large amount of RNA that is transported to the cytoplasm where it participates in protein synthesis. Newer and more sophisticated biochemical procedures have borne out this theory.

Another function ascribed to the nucleolus and resulting from microbeam irradiation is its role in cell division. The evidence here is conflicting. For example, Gaulden and Perry (1958) and Perry (1961) demonstrated that heterochromatic UV nucleolar irradiation from mid-telophase to mid-prophase results in mitotic delay, but

Cells irradiated in designated areas—incubated ^{3}H-cytidine, 8 hr.

Area irradiated	Mean grains per nucleolus	Number of cells	Area irradiated	Mean grains per nucleus	Number of cells	Area irradiated	Mean grains per cytoplasm	Number of cells
None	6.5	28	None	41	28	None	43	28
Nucleus			Nucleus			Nucleus		
one	6.2	15	one	36	15		35	26
two	4.8	12	two	29	12			
Nucleoli			Nucleoli			Nucleoli		
two	1.1	20	some	26	8	some	12	8
three	0.4	6	all (two)	20	15	all (two)	5.5	14
			all (three)	10	6	all (three)	3	6

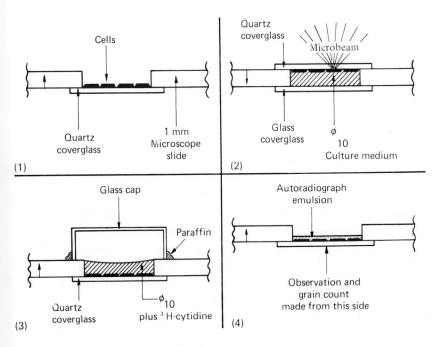

Fig. 5-14. Results of UV microbeam experiment where nucleoli and nonnucleolar regions of the nucleus were irradiated, and subsequent RNA synthesis assayed autoradiographically. (a) Tabulated data. (b) Sequence of the treatment of cells in a microbeam experiment (1), growth; (2), irradiation; (3), incubation; (4), autoradiography. (From Perry et al., 1961.)

(a)   (b)

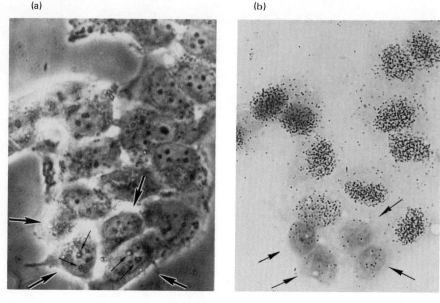

Fig. 5-15. CMP tissue culture cells following argon laser microir-radiation of nucleoli. (a) Live cells immediately following irradiation of nucleoli in four cells (large arrows); small arrows show irradiated nucleoli with lesions. (b) Autoradiograph of same cells following incubation in ^{3}H-uridine. (From Berns et al, 1969.)

small doses at 2804 Å produce mitotic acceleration. Sakharov and Voronkova (1965) irradiated nucleoli of swine kidney tissue culture cells with 2800 Å radiation and found a mitotic inhibition based upon the radiation dose rather than the radiation site (i.e., nucleolus). In even more recent studies by Berns and Cheng (1972) and Ohunki et al. (1972), employing argon laser microirradiation of nucleoli and nucleolar-associated chromosomes, it was found that both the stages of mitosis that irradiation occurred in, the site of irradiation (whether nucleolar-associated chromosomes, nonnucleolar chromosomes, or nucleolus), the radiation dose, and even the type of cells irradiated, were important in determining the response of the cells. For example, identical experiments conducted on an estab-lished cell line of the rat kangaroo and cells from primary cultures of salamander lung epithelium indicated that the salamander cells were generally more susceptible to mitotic inhibition (Fig. 5-16). Both cell types became less susceptible to mitotic inhibition as prophase progressed, but the salamander cells were much more susceptible to mitotic blockage in early stages than the rat kangaroo cells.

Prophase Irradiation: salamander (S) and rat kangaroo (K)

Prophase stage	Type of irradiation	Mitosis blocked		Mitosis continued		Percent continued	
		(S)	(K)	(S)	(K)	(S)	(K)
Early	All juxtanucleolar regions	8	7	0	1	(0)	(12)
Early	Juxtanucleolar of 1 nucleolus	6	12	0	10	(0)	(45)
Early	Random chromosomes (non-juxtanucleolar)	9	7	2	13	(18)	(65)
Early	No irradiation control	3	5	6	25	(67)	(84)
Middle	Juxtanucleolar of 1 nucleolus	15	–	4	–	(21)	–
Late	All juxtanucleolar regions	1	1	8	16	(89)	(94)
Late	Juxtanucleolar of 1 nucleolus	4	2	14	7	(78)	(77)
Late	Random non-juxtanucleolar chromosomes	1	2	4	8	(80)	(80)
Late	No irradiation control	1	0	4	5	(80)	(100)

Fig. 5-16. Comparison of laser radiosensitivity during mitosis in the salamander (*Taricha*) and the rat kangaroo (*Potorous*). (Modified from Berns and Cheng, 1972.)

A third possible function of the nucleolus is its role in the transport of nuclear informational RNA (messenger) from the nucleus to the cytoplasm. These studies were conducted by Henry Harris and have been discussed in the previous section.

Perhaps a fourth function of the nucleolus exists, and that is the role in its own formation. Since the nucleolus disappears visually during mitosis and is reconstituted in late telophase and early interphase, the question of its reconstruction is raised. Sakharov and Voronkova (1966) found that UV microirradiation of nucleoli in prophase resulted in abnormal nucleolar fragments being formed in late telophase. Similarly, they also found nucleolar fragments formed when they irradiated chromosomes in late prophase, metaphase, and anaphase. From these studies they concluded that nucleolar reformation was controlled by the nucleolus prior to its disappearance in prophase, and by the chromosomes after nucleolar disappearance had occurred. This theory is not out of line with the following evidence: (1) the nucleolus is organized by a chromosome region (the nucleolar organizer) that probably contains the ribosomal genes; and (2) some nucleolar constituents synthesized just prior to nucleolar disappearance are stored in the cell and re-used in the synthesis of the new nucleolus following mitosis. Biochemical evidence for this second point has been described recently by Phillips (1972), and argon laser microirradiation of the nucleolar organizer of mitotic chromosomes substantiates the first point. These studies will be discussed in the next section.

A more complicated study involving nucleolar irradiation has been conducted by Moore and Berns (1973). These studies were designed to elucidate the functional nature of the fine structural components of the nucleolus; the fibrillar and granular components. It is a well established fact that in proper concentrations the drug actinomycin D causes the *segregation* or separation of the fibrillar and granular components in tissue culture cells. When the drug is removed the nucleolar components reassociate into a normal nucleolus. Since nucleoli can be sensitized to argon laser light with quinacrine hydrochloride (Berns et al., 1969), it was possible to first segregate the fibrillar from the granular components with actino-mycin D, then sensitize with quinacrine hydrochloride, and finally selectively irradiate either of these components with the laser microbeam. The actinomycin and quinacrine were then removed and H^3-uridine, an RNA precursor, added to the culture medium for varying periods. The results of these studies indicated that general RNA synthesis was greatly reduced when the *light* granular components were destroyed, and only slightly affected when the darker fibrillar components were irradiated. These results bear out earlier suggestions that the light granular components contain the ribosomal DNA. Destruction of this material should greatly reduce the ability of the nucleolus to synthesize RNA. The fibrillar material was proposed to contain nucleolar products (ribosomes) which are continually turning over. Destruction of these components should only temporarily affect RNA synthesis by the nucleolus. This theory is borne out by the laser studies.

C. PARTIAL NUCLEAR
IRRADIATION–CHROMOSOMES

Perhaps the greatest number of microbeam experiments performed on one organelle has been on the chromosomes. These experiments can be divided into two general types of investigations: (1) those designed primarily for structural alteration of the chromosomes; and (2) those where primary intent is the study of chromosomal function and behavior.

In the mid 1950's Zirkle, Bloom, Uretz, and Perry published a series of papers describing the phenomenon of *chromosome paling* that resulted from exposing parts of individual newt *(Triturus viridescens)* chromosomes to UV and proton microbeams. Apparently the index of refraction of the irradiated chromosome region

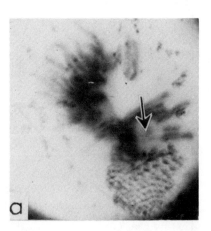

Fig. 5-17. Paling following UV microirradiation of *Triturus* mitotic chromosome. (From Perry, 1957).

was lowered. The result was a phase-light or *pale* spot when viewed through the phase-contrast microscope (Fig. 5-17). Subsequent Feulgen staining demonstrated a decrease in DNA content of the paled region. Alfert-Geschwind staining for basic (histone) protein demonstrated a reduced staining in the paling region and haematoxylin phosphotungstic acid staining indicated that there was some material still in the irradiated segment. The loss of DNA was indicated further by Perry's (1957) demonstration of a reduction in UV absorption at 2600 Å (nucleic acid absorption peak) in the paled chromosome region. However, the action spectrum (Fig. 5-18) of the paling phenomenon indicated that DNA was not the constituent initially absorbing the UV microirradiation (Zirkle and Uretz, 1963). The maximum paling efficiency was between 2700 Å and 2800 Å. This suggested that protein was the component absorbing the UV light and that DNA was affected secondarily. The sum total of these studies led Zirkle and Uretz (1963) to conclude that paling was

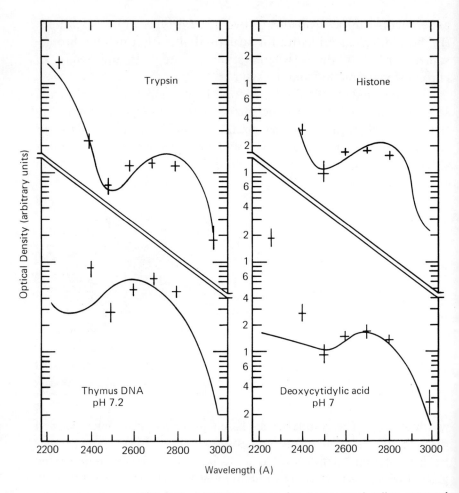

Fig. 5-18. Action spectrum for chromosomal paling compared
with absorption spectra for trypsin, DNA, histone, and deoxcytitdy-
lic acid. In each of the four charts, the seven experimental points
have had their ordinates multiplied by a convenient constant, and
the optical density of each substance is multiplied by an arbitrary
constant that brings the continuous absorption curve into best visual
fit with the seven points of the action spectrum. (Modified from
Zirkle and Uretz, 1963.)

due to a loss of both protein and DNA in a combination which is
detached from the chromosome by the UV photons that were
initially absorbed by the protein constituent (presumably histone).
Electron microscope analysis of the paled chromosome segments

demonstrated a loss in constituent A—a material possibly corresponding to DNA and histones—and the continued presence of a supportive constituent. (See detailed discussion in Chapter 7.) The term *DNA-steresis* was introduced to refer to the loss of chromosome material.

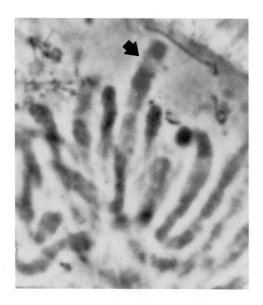

Fig. 5-19. A typical argon laser paling lesion on a metaphase salamander chromosome.

UV paling was produced in all mitotic stages from early prophase to late anaphase with the same levels of UV energy density. The degree of the paling appeared to depend upon the fraction of the total chromosomal material bombarded. For example, the paling spot from a 16μ diameter UV spot was much less intense than the paling spot produced with an 8μ spot, even though the energy density per unit area was the same. If the entire chromosome set or the entire cell was irradiated, practically no paling was detected. Another phenomenon was a spreading of the paling down the length of the chromosome for up to 20-30 min. postirradiation. The final paled segment was often 2-3 times the diameter of the initial irradiated spot. Both the spreading effect and the apparent dependence of paling intensity on irradiated area, rather than energy density, are phenomena that have not been adequately interpreted in the literature.

Chromosome paling appeared again in the literature when Berns et al. (1969) irradiated mitotic chromosomes of the salamander *Taricha* with blue-green light from an argon laser microbeam (Fig. 5-19). A spot diameter of 0.5-1 μm was employed and the energy in the spot was 1-25 μJ. The chromosomes were sensitized to the laser light with 5 min incubation in acridine orange (25μg/ml). Subsequent DNA-feulgen and protein (alkaline fast green) staining was negative in the paled area (Berns et al., 1971). The cells completed mitosis and were followed in culture for several days. Subsequent mitosis did not occur. However, when lower acridine orange concentration (0.1-0.01 μg/ml) and slightly higher energy densities (25-50 μJ) were used, chromosome paling that stained negative for DNA and positive for protein was observed. These studies were performed on salamander lung epithelial cells and two established cell lines of the rat kangaroo (*Potorous tridactylis*). The salamander cells were observed entering a subsequent mitosis, but they failed to complete division normally. In similar experiments on rat kangaroo cells, the cells have been followed repeatedly through subsequent mitosis (Fig. 5-20). These were the first observatiobs proving that cells with microirradiated chromosomes were capable of undergoing additional divisions (Berns et al., 1971).

In the preceding experiments the chromosome damage was attributed to energy absorption by the acridine orange which àttaches between the DNA base pairs. With the highest dye concentration, the amount of energy absorbed was probably great enough to damage both the DNA and protein. With the lower dye concentrations the damage was restricted to the DNA.

Subsequent to these studies, chromosomes were exposed to even greater argon laser energy densities (300-500μJ or 10^5 w/cm^2), and the result was a chromosome paling without any dye sensitization. Cytochemical staining indicated that the paled chromosome segment was positive for DNA and negative for histone protein (Berns and Floyd, 1971). By giving five to seven repeated irradiations at this energy level within a 10 sec period it was possible to alter both DNA and histone (Basehoar and Berns, 1973). These irradiated cells have been isolated and cloned into viable populations. Both these results and the preceding result indicate that selective alteration of

(a)

(b)

(c)

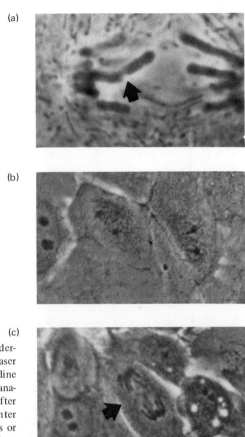

Fig. 5-20. Rat kangaroo cells undergoing subsequent mitosis following laser microirradiation (arrow) or acridine orange sensitized chromosomes: (a) anaphase chromosomes immediately after laser microirradiation; (b) two daughter cells 1 hr post-irradiation; (c) mitosis or irradiated cell 25 hr post-irradiation. (From Berns et al., 1971).

DNA, histone, or both components, is possible by choosing appropriate combinations of laser energy and acridine orange (Fig. 5-21). These results also were confirmed by using a functional genetic assay for DNA actitivy: the ability of the nucleolar organizer to construct a nucleolus.

The use of an organic dye laser microbeam employing visible laser wavelengths from the blue (4400 Å) to the red (6400 Å) produced chromosome paling that was more intense and more rapidly appearing than any of the preceding, non-acridine orange paling lesions. The paling was produced only by blue wavelengths. Power densities as high as 10^7-10^9 w/cm^2 were employed for all wavelengths (blue-red). These alterations have not yet been examin-

Altered chromosome component	Concentration of acridine orange μgm/ml	Energy $\mu J/\mu^2$	Feulgen stain	Protein stain	Nucleolar organizer function
DNA	1-0.01	25-50	–	+	–
protein (histone)	0	300-500	+	–	+
DNA and protein	1	100-500	–	–	–
	0	1000	–	–	–

Fig. 5-21. Summary of chromosome components affected by various combinations of argon laser energies and acridine orange concentrations.

ed cytochemically or ultrastructurally. In all these studies (argon and dye laser), cells finished mitosis and were followed in culture for several days.

Functional inactivation of specific chromosomal regions have fallen into two classes of experiments: UV inactivation of kinetochores, and laser inactivation of the nucleolar organizer.

Kinetochore inactivation was accomplished by Uretz et al. (1954) using a UV microbeam and by Bloom et al. (1955) with a proton microbeam. The sum total of these experiments was a consequent inability of the chromosomes to move in a directed fashion towards the metaphase plate after kinetochore irradiations. However, in a later study by Bajer and Bajer (1961) employing the Zirkle UV microbeam to irradiate kinetochores of mitotic cells of the water lily, *Haemanthus*, chromosome movements to the metaphase plate were not affected by kinetochore inactivation.

In their extensive series of experiments on *Haemanthus* chromosomes, Bajer and Bajer (1961) described both indirect and direct effects in the irradiated mitotic cell; over several hundred kinetochores and chromosome arms were irradiated with a 2-4 μ UV microbeam. Direct effects such as chromosome paling, chromosome *stickiness*, and swelling and spreading of the irradiated region were observed. Indirect cellular effects such as irregular anaphases, formation of restitution nuclei, prolonged mitosis, and cell death were often detected with the longer and more intense radiation doses. The most striking result of these experiments was the observation that anaphase-irradiated chromosomes tend to change their behavior and act like prometaphase chromosomes—e.g., move towards the equatorial or metaphase plate. It is theorized that this

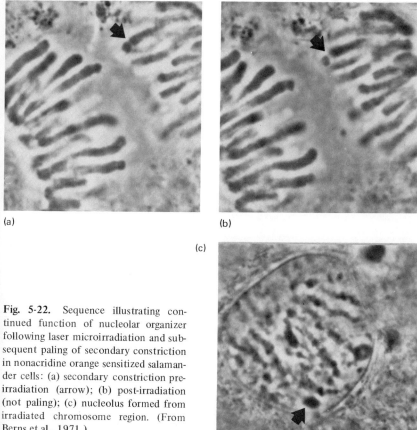

Fig. 5-22. Sequence illustrating continued function of nucleolar organizer following laser microirradiation and subsequent paling of secondary constriction in nonacridine orange sensitized salamander cells: (a) secondary constriction pre-irradiation (arrow); (b) post-irradiation (not paling); (c) nucleolus formed from irradiated chromosome region. (From Berns et al., 1971.)

result is due to the unmasking of a second kinetochore by the UV irradiation, which removes a protective-repressing material. The result is an initiation of a second spindle fiber attachment to the kinetochore, which pulls the chromosome in a direction opposite to the first fiber. Under two equal forces the chromosome is pulled towards the equatorial plate.

A number of investigations designed to elucidate the location and functional role of the nucleolar organizer, which is the secondary constriction region, has been undertaken using the argon laser microbeam. Mitotic cells of the rat kangaroo cell lines and salamander lung explants have been irradiated with and without acridine orange sensitization. Irradiation using single laser pulses with up to 500 μJ in the focal spot have produced chromosome

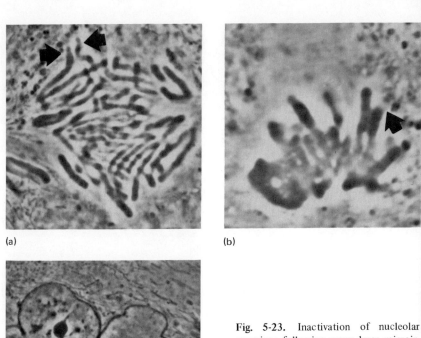

(a)

(b)

Fig. 5-23. Inactivation of nucleolar organizer following argon laser microirradiation of secondary constriction of acridine orange treated salamander cells: (a) pre-irradiation (arrows indicate secondary constriction; (b) post-irradiation of one secondary constriction (arrow indicates paling spot); (c) two nuclei several hours post-irradiation (arrow indicates site lacking third nucleolus). (From Berns et al., 1970.)

(c)

paling, as described earlier, but have not resulted in inactivation of the nucleolar organizer (Fig. 5-22). This result further confirmed that nondye sensitized chromosome paling does not involve DNA damage. When acridine orange was employed to sensitize the chromosomes, irradiation of the secondary constriction region of the anaphase chromosomes resulted in the inability of that particular chromosome to synthesize a nucleolus (Fig. 5-23). Initial studies on salamander anaphase chromosomes involved irradiation of the entire secondary constriction with a small chromosome area on each side of the constriction. Cells were formed lacking either one or two of their normal number of three nucleoli. It was observed frequently that the remaining one or two nucleoli were considerably larger than

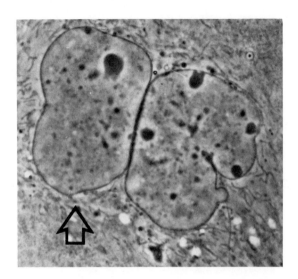

Fig. 5-24. Large compensatory nucleolus formed in salamander cell nucleus (arrow) following experimental reduction of nucleolar number. This nucleus should normally contain three or four nucleoli as in the adjacent nucleus. (From Berns et al., 1970.)

normal—suggesting a nucleolar compensation mechanism (Fig. 5-24). Similar studies on the male rat kangaroo cell line, which normally has only one nucleolar organizer and, therefore, one nucleolus, resulted in the production of several small micronucleoli following inactivation of the major nucleolar organizer (Fig. 5-25). These results suggested either the existence of secondary nucleolar organizers, or the storage and repopulation of new nucleoli with constituents of the old nucleolus.

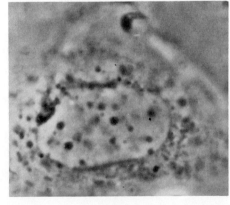

Fig. 5-25. Numerous small micronucleoli in nucleus of the male rat kangaroo (PTK$_2$ cell line) following laser microirradiation of the single major nucleolar organizer. (From Berns et al., 1972.)

Refinement of the argon laser microbeam to produce lesions of less than 0.5μ permitted the fine dissection of the nucleolar organizer (Fig. 5-26).   Irradiation just adjacent to the secondary constriction resulted in a consistent loss of nucleolar organizing capacity.   These experiments suggest that there may be some validity to the long hypothesized and often ignored theory that the nucleolar organizer is adjacent to the secondary constriction region and not within it.

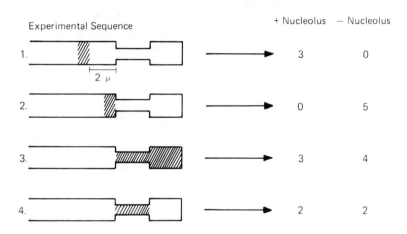

Fig. 5-26.   Summary diagram.   Fine dissection of the nucleolar organizer region of salamander chromosomes with an argon laser microbeam.   The portion of the chromosome irradiated is indicated by shading.   Sequence 1 involved irradiation of a region up to 2 μ down from the constriction.   Sequence 2 involved irradiation of the chromosome region immediately adjacent to the constriction. Sequence 3 involved irradiation of both the constriction and the distal satellited top.   Sequence 4 involved irradiation of the constriction alone.   (From Berns and Cheng, 1971.)

A somewhat different, but similar nucleolar organizer irradiation experiment involved laser microirradiation of the prophase chromosome regions that were associated with the nucleolus (Fig. 5-27).   Irradiation of the juxtanucleolar chromosome regions resulted in either a decrease in nucleolar number, as predicted from the preceding experiments, or a complete reversion of the cell back into interphase.   This latter result led to a series of studies designed to elucidate the effect of microirradation of prophase chromosomes on the mitotic process.   As mentioned previously in this chapter, similar

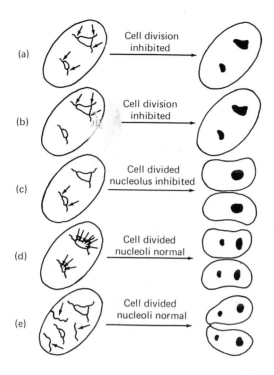

Fig. 5-27. Summary diagram of argon laser microirradiation of nucleolar-associated chromosomes and their role in nucleolar organization and mitosis. Arrows indicate the sites of irradiation. (a), (b) cell division was inhibited; (c) cell divided but formation of one nucleolus was inhibited; (d), (f) cell divided and same number of nucleoli were formed as were present in the mother cell. (From Ohnuki et al., 1972)

studies were performed on salamander and rat kangaroo cells. In both cell types mitosis was significantly blocked when irradiation occurred in early prophase, regardless of whether or not the irradiated chromosomes were nucleolar-associated. It was also determined that with identical irradiation conditions, the salamander cells were more susceptible to mitotic inhibition than the kangaroo cells (Fig. 5-16). All these studies were conducted using high laser energies (100 $\mu J/\mu^2$). Apparently the site-specific inhibition of mitosis and reduced nucleolar number holds only for the lower energy levels (10-20 $\mu J/\mu^2$).

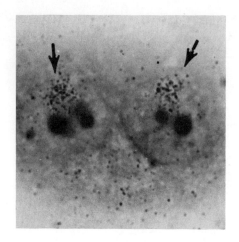

Fig. 5-28.   Incorporation of DNA precursor (^{3}H-TdR) following UV microirradiation of a 5 μ area of the nucleoplasm during G_1 or G_2. Note unscheduled DNA synthesis at irradiation sites (arrows). (From G. Moreno, 1971.)

Another interesting study involving UV microirradiation of nucleoplasm is worthy of mention in this section because the material affected was undoubtedly chromosomal.   Moreno (1971) demonstrated that when a 5 μ classical UV (10^{-1} erg/μm^2) beam was focused into the nucleoplasm of KB liver cells in culture, there was an unscheduled uptake of H^3-TdR, a DNA precursor, in the irradiation region (Fig. 5-28).   However, when the irradiation and subsequent incubation in the isotope occurred during the normal DNA synthesis phase of the cell cycle, there was a reduced amount of isotope incorporation in the irradiated region.

D.   PARTIAL CYTOPLASMIC IRRADIATION–
MITOTIC SPINDLE

Many of the studies involving the inhibition of the mitotic spindle have employed microirradiation of cytoplasmic constituents.   In the mid 1950s Bloom, Zirkle, and Uretz demonstrated that the UV irradiation of the cytoplasm of newt cells either inhibited the formation or destroyed the spindle.   Surprisingly, however, the cells still underwent a mitotic-like process.   Within 20 min after irradiation of a metaphase cell, the spindle disappeared, and soon after the chromosomes collapsed into a haphazard pattern.   Within about 60 min an orderly chromosome pattern resembling a *rosette* was achieved.   Several hours after irradiation the rosette separated into

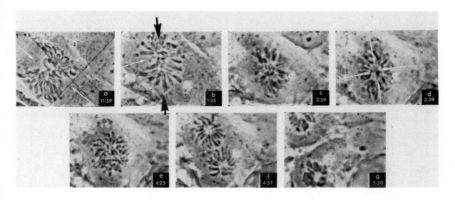

Fig. 5-29. Quasirosette formed following heterochromatic UV irradiation of spindle area with heterochromatic UV radiation. (a) metaphase cell pre-irradiation, target area in cross hair rectangle; (b) post-irradiation; (c) disorganization and haphazard chromosome arrangement; (d) quasirosette; (e), (f), (g) separation of quasirosettes. (From Zirkle, personal communication.)

two groups of chromosomes, each one being separated into a daughter cell. (See Fig. 5-29 for sequence of these events.) The authors termed this a *false anaphase* and pointed out that even though the chromosomes were equally divided between the two cells, normal separation of chromatids did not occur. If the cytoplasm of mitotic cells was irradiated in prophase before spindle formation had occurred, a spindle was never detected and the chromosomes did not form a metaphase configuration. Instead, they went directly into the rosette pattern and divided as in the previous example. The authors performed a UV action spectrum of spindle disappearance and consequent mitotic delay. They demonstrated that the effects occurred, regardless of the wavelength, between 2250 Å and 3000 Å, but most efficiently at 2250 Å; thus closely resembling the absorption spectrum of tyrosine-rich protein. Later studies by Wada and Izutsu (1961a, b) on mitotic *Tradescentia* hair cells demonstrated that UV microbeam irradiation of the early polar cap caused spindle disappearance and the reconstitution of a nucleus. If irradiation occurred later, chromosome stickiness occurred, the two groups of sister chromatids were pushed towards one pole, and a phragmoplast formed causing the production of a binucleate and an anucleate cell.

Direct irradiation of the spindle itself was effective also in causing its own disappearance. Such studies by Davis et al. (1957) were done with a polonium alpha microbeam in chick fibroblasts, and with a heterochromatic UV microbeam in *Tradescentia* by Izutsu (1961).

I have mentioned only a few of the microbeam studies involved with spindle effects. A wide variety of other organisms have been studied: *Haemanthus* (water lily) (Bajer and Bajer, 1961); the embryos of *Cecidomycidae* (a dipteran fly) (Geyer-Duszynska, 1959, 1961); and leucocytes from newt liver (Amenta, 1962). In general, the results of all of these studies pointed to the importance of the spindle in organizing a normal mitosis and led to the hypothesis that cytoplasmic irradiation produces a nuclear or spindle *poison* presumably by a photochemical process. What is perhaps the most remarkable result from these studies is the demonstration that even though the spindle is important to a normal ordered mitotic event, the necessity for mitosis is so programmed within the cell, that even when the spindle is damaged or destroyed, a mitotic-like separation of the chromosomes still occurs. One can only speculate whether this is a built-in evolutionary plasticity or merely the result of a set chain of events that, once mechanically set in motion, must be carried to completion.

E.   PARTIAL CYTOPLASMIC IRRADIATION— MITOCHONDRIA

Selective mitochondrial alteration has been performed exclusively with laser microbeams. Amy and Storb (1965) first described the destruction and/or alteration of mitochondria in KB and fibroblast tissue culture cells supravitally stained with Janus Green B after cytoplasmic irradiation with the 5-8μ diameter spot of a ruby laser. Extensive studies were conducted to describe the morphological alteration at the light and electron microscope levels (Storb et al., 1966). Two general classifications of damage were observed: primary, occurring at the immediate site of the irradiation and attributed directly to the radiation; and secondary damage outside the site of irradiation but also related to the irradiation. The primary lesions were further subdivided into three categories: *light, moderate,* and *heavy*. At the ultrastructural level the three types of primary lesions were distinct. The light lesion exhibited an electron-dense material in the mitochondrial matrix, and the moderate alteration had an electron-opaque region surrounded by membranous residual mitochondrial material. In both the light and moderate alterations only the mitochondria in the irradiated cytoplasm were affected. The heavy lesions appeared as electron-dense masses in the irradiated portion of the cytoplasm. Mitochondrial structures were not detected in the damaged area, and other organelles in the irradiated region were damaged.

Cell viability studies (Amy et al., 1967) indicated general reduction in mitotic capability for the least severe alterations (6 degrees of primary damage were described) and a complete inability to undergo mitosis for all other irradiated cells. Biochemical analysis of the KB cells using H^3-uridine, an RNA precursor, suggested either a reduction in RNA synthesis or a dilution of the isotope pool by release of nonisotopic RNA precursors (Wertz et al., 1967). Extensive cytochemical analysis of the mitochondrial dehydrogenase enzymes demonstrated a consistent reduction of enzyme activity only in the case of heavy primary damage (Storb et al., 1966).

Subsequent to the above studies, Berns and Osial (unpublished data) irradiated the large mitochondria of rat heart cells with a ruby microbeam and demonstrated considerable morphological alteration without vital staining. This is not in agreement with the earlier studies where Janus Green B was required to sensitize mitochondria to the ruby laser. However, it is possible that the large concentration of cytochromes per unit area in the large heart mitochondria was sufficient to absorb enough laser energy for damage production.

The blue-green argon laser microbeam has been employed extensively by Berns et al. (1970, 1972) for irradiation of the large (1-5 μm in diameter) rat myocardial cell mitochondria. It has been suggested that cytochrome C acts as a natural chromophore absorbing the laser light. Cells have been examined morphologically with the phase and electron microscope (Adkisson et al., in preparation), cytochemically for enzyme activity, and the functionality of the cells has been assayed by analysis of contractile activity (Berns et al., 1972).

The phase microscope alterations have been designated *light, moderate,* or *severe.* The light alterations appear generally as a small dark spot at the focal point of the irradiation with a paling of the

(a)   (b)

(c)

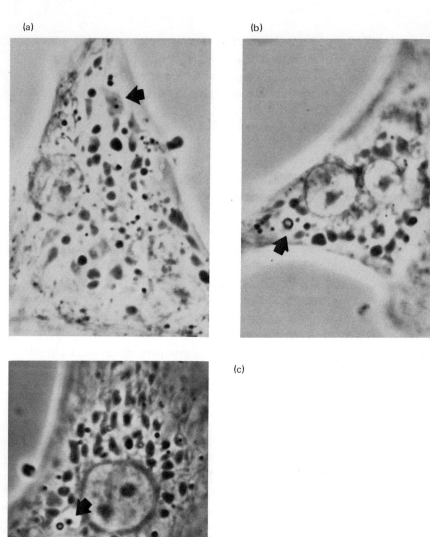

Fig. 5-30.   Three types of mitochondrial lesions in rat heart myocardial cells following laser microirradiation, (a) light, (b) moderate, (c) severe. See text for discussion.

(a)   (b)

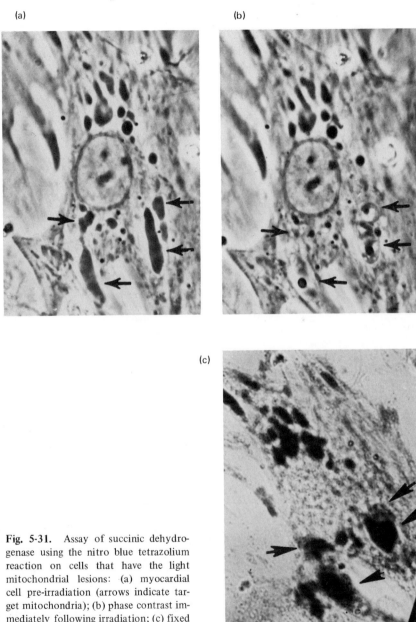

(c)

Fig. 5-31. Assay of succinic dehydrogenase using the nitro blue tetrazolium reaction on cells that have the light mitochondrial lesions: (a) myocardial cell pre-irradiation (arrows indicate target mitochondria); (b) phase contrast immediately following irradiation; (c) fixed and stained preparation. Note the high SDH levels in the irradiated organelle.

rest of the mitochondrion (Fig. 5-30a). The moderate lesion appears as a hole or phase-light spot at the site of irradiation, and the unirradiated portion of the mitochondrion maintains its optical phase density (Fig. 5-30b). The severe alteration appears to be complete disruption of the irradiated mitochondrion (Fig. 5-30c). Despite the relative ease with which the alterations could be

Group	Definition	Subgroup	Time from irradiation (min)			No. cell
			0*	3	30	
I	No change	–	–	–	–	14
II	Change followed by	A	Stop	Fibrillation	Rhythmic	9
	return to rhythmic	B	Fibrillation	Rhythmic	–	5
	contraction	C	Stop	Irregular contractions	Rhythmic	1
		D	Stop	Rhythmic	–	1
		E	Irregular	Rhythmic	–	1
		F	No effect	Irregular	Rhythmic	1
III	Stops contractions	G	Fibrillation	Stop	–	2
		H	Stop	Fibrillation	Stop	2
IV	Death	I	Death	–	–	14
		J	Stop	Death	–	4
		K	Fibrillation	Death	–	2

*Time 0 was immediately after irradiation.

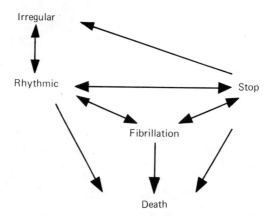

Fig. 5-32. Contractile responses of myocardial cells following argon laser microirradiation of a single mitochondrion. (a) tabular form; (b) flow diagram illustrating all observed pathways. (From Berns et al., 1972.)

classified with the light microscope, preliminary ultrastructural studies indicated that a variety of fine structural changes may be present for any given light microscope lesion type. However, one type of lesion was frequently detected in mitochondria that had the least severe lesion (see Chap. 7). Cytochemical staining for succinic dehydrogenase (Fig. 5-31) demonstrated a marked reduction of enzyme in the irradiated mitochondrion when the lesions were of the moderate or severe type. The light lesions did not result in a reduction of SDH activity. Unirradiated mitochondria in all cells stained normally for succinic dehydrogenase. These studies suggested microirradiation of single mitochondria does not significantly affect the other mitochondria in the cell—at least with respect to succinic dehydrogenase.

The specific lesion types also could be correlated with contractile responses. These contractile responses were divided into four general categories: no change, change followed by a return to rhythmic contraction, cessation of contraction, and death. Cells were observed going through many different sequences of contractile change, and these are summarized in Fig. 5-32. Of a particular interest were the return of cells to a rhythmic contractile rate identical to the pre-irradiation rate and the frequent induction of a single cell fibrillation. Both of these functional responses are under intense investigation because of their immediate relevance to general heart physiology and the regulation of myocardial cell activity.

It was also possible to correlate the lesion type with contractile response (Fig. 5-33). As one might predict, the severe lesions frequently resulted in cell death, and the light lesions frequently did

Functional response	Weak I	Lesion type Moderate II	Severe III	Totals
No change – group I	8	6	0	14
Return to rhythmic – group II	0	16	2	18
Stop – group III	0	4	0	4
Death – group IV	0	2	18	20
Totals	8	28	20	

Fig. 5-33.   Correlation of functional response and lesion type in rat myocardial cells following argon laser microirradiation of single mitochondria. (From Berns et al., 1972.)

not elicit contractile changes. The moderate lesions produced the contractile changes in which the cells returned to a normal contractile rate.

The direct cause of the morphological and functional responses are not known, but several possibilities have been suggested: (1) indirect cell membrane damage; (2) release of Ca^{++} from the mitochondria; (3) thermo-mechanical effects on the cytoplasm; and (4) conversion of laser energy to ATP. With respect to the last possibility, Salet (1972) has described an increase in contractile rate following mitochondrial irradiation with green laser light from a frequency doubled neodymium glass laser. He suggests the direct conversion of laser energy to chemical energy in the form of ATP as a possible mechanism.

Organisms, Gametes, and Embryos
SIX

A. UNICELLULAR ORGANISMS

Protozoans have been favorite targets of microbeamists. *Amoeba* first were irradiated by Tchakhotine in 1935, in the 1960's by Jagger et al. (see discussion in Chap. 5, p. 64, Bessis et al. (1965), and Saks et al. (1965). In his early studies using a 2800 Å

Fig. 6-1. Pinching off of amoeba cytoplasm following repetitive irradiation with a ruby laser microbeam of 100 mJ. (From Saks et al., 1965.)

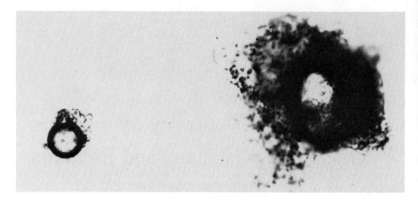

Fig. 6-2.   Ejection of portion of oil droplet with adhering cytoplasm following irradiation of oil droplet in cell cytoplasm. (From Saks et al., 1965.)

microbeam, Tchakhotine demonstrated that when an elongating pseudopod was irradiated, cytoplasmic streaming stopped and then started again in the opposite direction.   However, when a pseudopod of the mold *Actinospharium* was irradiated, the surrounding pseudopodia actually changed their direction and moved toward the damaged area.   Using a ruby laser microbeam, Saks et al. (1965) demonstrated that following nuclear irradiation with 100 mJ of energy, cytoplasmic streaming stopped and then resumed, either in the same or different directions.   They demonstrated also that amoebas with damaged cytoplasm, actually ejected the damaged material and moved away from it (Fig. 6-1).   A similar reaction was observed when oil droplets that were injected into amoeba cytoplasm were irradiated (Fig. 6-2).   The same authors also demonstrated that either nuclear or cytoplasmic irradiation was effective in substantially reducing the growth rates of amoebas (Fig. 6-3).

The biology of *Paramecium* has been studied effectively with microbeams.   When Tchakhotine (1935) UV-irradiated one of the two contractile vacuoles, it swelled, stopped contracting, and the unirradiated vacuole started to contract at an accelerated rate. The irradiated vacuole usually recovered after one day.   When the two vacuoles were irradiated, both swelled, stopped contracting, and the cell usually died.   However, in a few cases two new vacuoles were formed.   When this occurred, the cell survived, and the old vacuoles eventually recovered.   In another protozoan, *Euplotes*, the damaged

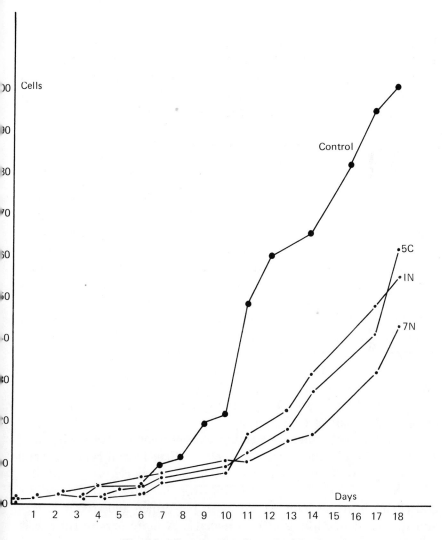

Fig. 6-3. Growth rates of amoeba following ruby laser irradiation compared to control unirradiated cells; 5c, five repetitive pulses to the nucleus; 7N, seven repetitive pulses to the nucleus; 1N, one nuclear pulse. (From Saks et al., 1965.)

vacuole was eliminated from the cell and a new vacuole formed. Both series of experiments on *Paramecium* and *Euplotes* indicated that there are definite built-in biochemical regulatory mechanisms for the control and differentiation of contractile vacoles. Other morphogenetic studies on *Euplotes* conducted by Wise (1965) were designed to investigate the repair and regeneration of primordial and differentiated structures, especially ciliary and cortical organelles. These studies employed a heterochromatic UV microbeam with a spot diameter of 12 μ. Shimomura (1967) also used a 2 μ heterochromatic ultraviolet microbeam to study the relationship between stomatum (mouth) formation and mitosis in *Euplotes.* Cells were irradiated at various times of the cell cycle in the region that would give rise to the new stomata. Division was inhibited when irradiation occurred prior to the initiation of stomatum formation, but once this process began, irradiation no longer inhibited mitosis. Whether or not stomatogenesis has a direct relationship to mitosis is still somewhat unclear, but the results re-emphasize clearly, that once a given point in time is reached (probably a physiological and biochemical state) and mitosis commences, the cell is programmed into a process that is very difficult to reverse.

In a later series of studies, Hanson (see Moreno's 1969 review) studied the morphogenesis and regeneration of the anal structures of *Paramecium*, and Franckel studied the oral structures of the ciliate *Glaucoma.* In both series of experiments a heterochromatic UV microbeam was used. The general aim of the experiments was to damage or destroy specific areas of the oral structures and study consequent repair or regeneration.

A rather voluminous literature exists on the microirradiation of a variety of other specialized organelles. Flagella and cilia of several organisms have been UV-microirradiated: *Paramecium, Frontonia, Climecostomu,* and *Stentor.* Studies on the growth, regeneration, and ciliary beat mechanism have been performed and are referenced and discussed in Zirkle's (1957) review. A more precise ruby laser microbeam study on the propagation of flagellar beat activity was performed on sperm tails (see discussion later in this chapter).

A rather unusual organelle, the peduncle of *Vorticella,* was UV-microirradiated by Tchakhotine (1936). This long organelle attaches the main body of the protozoan to the substrate, and consists of a central contractile filament, the myonema, and a transparent outer elastic sheath. By UV microirradiation of the cell body itself it was possible to elicit specific contractile responses of the myonema. The generation and transmission of the physiological

(a)

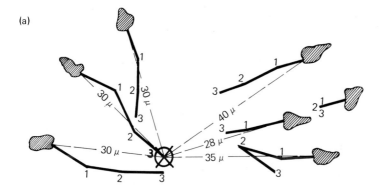

(b)

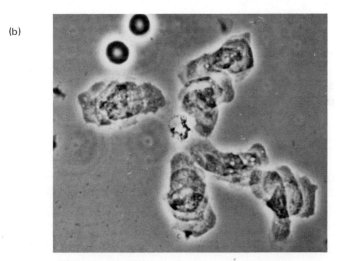

Fig. 6-4. Necrotaxis attraction of white blood cells to dead red blood cells. (a) A diagram of leukocyte movements after destruction of an erythrocyte by the laser. The hatched areas indicate the position of the leukocytes at the outset; the cross shows the site at which the red cell was destroyed; each number designates the minute-by-minute location of the leukocytes. Note that all the cells are attracted towards the target during the first minute but that they do not all continue in that direction. (b) Superimposed sequential photographs illustrating the course of leukocytes attracted by the red cell which has been destroyed by the laser. (Figures courtesy of Marcel Bessis, from *Living Blood Cells and their Ultrastructure,* Springer, 1973.)

impulses was not accompanied by visible morphological changes in the myonema. These particular studies indicate how the microbeam may be used to study a rather specialized and uncommon organelle.

Another specialized organelle that has been microirradiated with UV radiation is the eyespot of the protozoan *Euglena*. This organelle is known to function in phototactic behavior. Tchakhotine (1936) UV-irradiated this organelle and then demonstrated a loss of phototaxis. In a subsequent experiment he irradiated the eyespot with visible monochromatic light and demonstrated that immobile animals started to move when their eyespots were exposed to blue or blue-violet light, thus proving that the eyespot really was the effective site of energy absorption for the phototactic response.

The chloroplasts of a variety of plant and animal cells have been microirradiated with visible laser irradiation. In one of his early laser studies Bessis and TerPogossian (1965) microirradiated *Euglena* with a ruby laser. The energy absorption was so efficient because of the chlorophyll that the organism was killed instantly. An interesting response was that adjacent *Euglena* exhibited marked negative response and moved away from the dead organism. The opposite result was observed when a human red or white blood cell was destroyed by ruby laser microirradiation. Surrounding unirradiated white blood cells were attracted to the dead cell in a process termed *necrotaxis* (Fig. 6-4). Both of these studies demonstrated that cells release substances that either repel or attract other cells in a process often termed *chemotaxis*. The particular biological significance of these observations is that cells, indeed, do release small quantities of either attractant or repellent molecules, and that other cells are able to respond to these substances. The question of cell recognition, attraction, and aggregation is of fundamental interest in developmental biology.

In this section I have mentioned only a few of the numerous microbeam experiments that have been performed on unicellular organisms. Rather than merely listing all the studies, I have attempted to mention and discuss a sample that should illustrate the diversity of structures and processes studied. For more thorough treatment of these studies, and a more complete view of the other experiments, the reader should consult one of the major reviews or the original papers.

B. MULTICELLULAR ALGAE

The multicellular algae fall into two categories: (1) the colonial forms, and (2) the noncolonial forms. Of the first type, *Volvox* has been irradiated with an argon laser microbeam by W. Hand (Occidental College, California, unpublished data). Various cells in the colony were microirradiated, and the effect on the whole colony studied. Of particular interest were the interactions between cells in the colony.

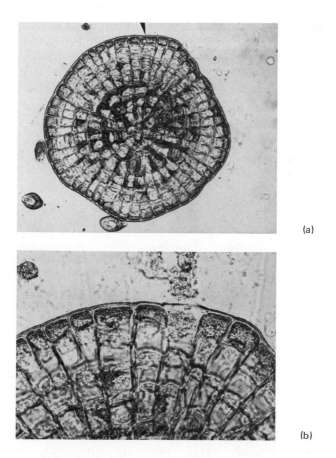

(a)

(b)

Fig. 6-5. Irradiation of chloroplast in a cell of the alga Coleochaete with blue laser light. (a) thallophyte with target cell indicated (arrow); (b) post-irradiation.

Of the second type, the noncolonial green algae, cells of the *Coleochaete* thallophyte were irradiated with both an argon and dye laser. The chloroplasts absórbed the laser energy and the result was either destruction of the entire cell (Fig. 6-5), or selective damage to the chloroplast. Ultrastructural examination of the irradiated chloroplasts demonstrated various degrees of damage to the chlorophyll-containing grana of the chloroplasts, depending upon the levels of energy employed. The ability to destroy single cells in a developing multicellular thallophyte has permitted detailed studies on morphogenesis and growth mechanisms. For example, when the specialized seta-bearing cells were destroyed in a young thallophyte, new ones differentiated, and the final number of the seta cells was brought up to the normal amount (Fig. 6-6). In addition,

Number Seta-Bearing Cells in Thallus

	Control thalli		Thalli with three seta-bearing cells irradiated		Thalli with three vegetative cells irradiated	
Initial, pre-irradiated	2	3	4	4	2	2
Initial, post-irradiated	2	3	1	1	2	2
Final no. seta cells	7	8	8	8	6	4
Final % seta cells	4.5%	3.9%	5.4%	3.7%	3.4%	2.6%

Fig. 6-6.   The development of setae-bearing cells in thalli that have had setae-bearing and vegetative cells irradiated with the argon laser. (From McBride and Berns, unpublished.)

Number of Cells in Thallus

			Treatments					
	A		B Thalli with three seta-bearing cells			C Thalli with three vegetative cells		
Time	Control		Average	irradiated	Average	irridiated		Average
Pre-irradiated	55	39	47	60	42	51	43	48	46
Post-irradiated 24 hr.	77	60	69	87	59	73	63	68	66
Post-irradiated 48 hr.	95	71	83	112	69	92	68	77	73
Post-irradiated 120 hr.	203	157	180	215	149	182	153	176	165
Total new cells	148	117	133	155	107	131	110	128	119

Fig. 6-7.   Total growth of coleochate thallus following argon laser microbeam irradiation of setae-bearing and vegetative cells. (From McBride and Berns, unpublished.)

destruction of cells in the inner portions of the thallophyte resulted in subsequent growth and mitosis of adjacent inner cells. Normally, elongation and division are confined to the marginal cells of the thallophyte (Fig. 6-7). These studies demonstrated built-in self-regulatory mechanisms for both seta cell differentiation and vegetative cell growth.

In later studies the same experimental approach was employed to study the process of dedifferentiation and subsequent redifferentiation of cells in the thallus. In those studies the non-setae bearing vegetative cells were all destroyed in a young thallophyte. Following this treatment the remaining seta cell appeared to lose its differentiated characteristics; the seta broke down, and the typical crescentic chloroplast of this cell became large and rectangular. Subsequently this dedifferentiated cell began to divide, giving rise to progeny that divided, and some of which became setae cells and others of which remained vegetative cells. This entire developmental sequence was repeatable and easily observed cytologically. Because of the beautiful cytology of the alga cells, the ease of growing the cultures, and the fact that the microbeam technique can be easily applied to this system, considerable insight into basic developmental mechanisms could be achieved.

Other studies on the green alga *Zygnema* were conducted many years ago (1942) by Petrova using a polonium alpha particle microbeam. In these studies the elongate alga was irradiated in either the nucleus or cytoplasm and consequent survival and division were recorded. It was found that division delay and mitotic inhibition were easily produced when the nucleus was bombarded, but that it required 700 times as much radiation to affect this response by cytoplasmic irradiation. Furthermore, when division delay was produced by cytoplasmic irradiation the cells eventually died, but in the case of nuclear irradiation the cells often recovered. These studies are particularly interesting in light of Jagger's more recent work on amoeba (Chap. 5, page 64).

C. GAMETES AND EMBRYOS

Partial irradiation of embryos and gametes is not always performed with the intent of studying the development of the organism. Often the gametes or early developing embryos provide excellent opportunities to study basic cellular processes. Therefore, this section will involve a discussion of partial embryonic and gametic irradiation for the purpose of studying basic cellular function, as well as development of the organism. A gamete is the haploid (one-half normal chromosome number) reproductive cell of the organism. These may be either sperms or eggs. Often these cells are cytologically ideal for microbeam irradiation. Mature sperm and immature spermatocytes have been irradiated to study such basic biological problems as chromosome movement during cell division, spindle fibers and mitotic traction, organization and packaging of chromosomal DNA, and flagellum activity.

The spermatocytes of the silkmoth *Bombyx* have very clear mitotic chromosomes whose movements have been studied with microcinematography following UV microirradiation. In 1964 and 1965 Nakanishi irradiated one chromosome group at the cell pole with a 5 micro UV microbeam and observed a rapid movement of the entire group to the opposite pole. Nuclear membrane formation was inhibited and cell division failed to occur, even though a cleavage furrow did appear. Control irradiation of nonchromosomal sites, likewise, inhibited cell division, but the unusual chromosome migration was not observed. The authors interpreted the unusual chromosome movements as resulting from the destruction of the connection between the chromosomes and the cell surface.

In a rather extensive series of studies on chromosome movements and spindle fiber association Forer (1965-1967) irradiated spindle fibers of primary spermatocytes of the crane fly (*Nephrotoma*) with a UV microbeam. He was able to cause a localized loss in fiber birefringence similar to that described in earlier studies by Inoué. By studying the subsequent movements of the negative birefringent region in relation to the unirradiated regions of the spindle fibers, and the chromosomes themselves, several hypotheses were made regarding the role of spindle fibers in the anaphase mitotic process: (1) the chromosomal fibers are the sites of force production or transmission; (2) chromosome movements are independent of chromosome fiber birefrigence; (3) chromosome movement is not usually related to chromosome fiber movement; (4) movement of sister dyads (double chromatids) are not independent; (5) different pairs of dyads in the same cell are independent; (6) normal chromosome movement requires an undamaged traction element extending for at least one half the fiber length from the chromosome; and (7) irradiation of interzonal spindle fibers produces mitotic delays, but irradiation of the half spindle produces delays similar to those observed with the cytoplasmic irradiation. When cells were irradiated in metaphase (instead of anaphase as described above): (1) chromosome movements depended upon intact spindle birefringence; (2) different pairs of dyads did not move independently; and (3) the irradiation in general had a greater effect on the production of a force mechanism.

The above studies by no means clear up the many questions concerning the precise role and mode of action of the spindle fibers in the mitotic process. However, they do suggest several rather significant facts, which may or may not be fully confirmed in the future. First, there is a site of force production, or transmission, at the level of the chromosome fibers; second, there are really two independent elements, one involving birefringence, and another with traction; and third, in anaphase the two elements are separate but in metaphase they are closely related.

Microirradiation with plane-polarized UV light has been employed extensively by Inoué and Sato (1967) to investigate the organization of DNA in sperm heads of the insect *Ceutophilus.* Because of the normal orientation of the helical DNA molecule in the sperm head, negative birefringence with an axis parallel to the geometrical axis can be demonstrated with the polarizing microscope. This negative birefringence can be abolished by UV microirradiation.

In mature spermatocytes, *microdomains* with axes of birefringence zigzagging regularly around the axis of birefringence of the sperm head were demonstrated. This suggested a helical molecular arrangement. By irradiation with a plane-polarized UV microbeam at an angle of 45° to the sperm head axis it was possible to reduce the birefringence of the microdomains unequally according to the orientation of their axes of birefringence. This led to a rotation of the mean optical axis of the sperm head. These values permitted the determination of the orientation of the optical axes of the microdomains. Similar studies on the sperm head of *Drosophila buskii* demonstrated analogous microdomains. Further refinements in the technique eventually permitted calculation of the optical axis orientations and determination of the birefringence patterns of each one of the irradiated and unirradiated microdomains at different angles to the sperm head axis. The results of these experiments suggested that the organization of DNA within the microdomains is variable, with the planes of the bases zigzagging around the mean axis of the microdomains. Extrapolating these findings, the authors proposed a model for DNA arrangement in the sperm head. Their model envisioned a 200 Å diameter filament, containing several DNA molecules with parallel axes, coiled into a double helix (super-coil). It was suggested that in the sperm head, alignment of segments composed of two such double helix super-coils is found with a common geometrical axis. These segments would be the chromosomes, each composed of two intertwined chromonemata. Such a model was also proposed by Sato and Muller (1966) for the sperm heads of *Loligo pealii* and was based on similar experimental results.

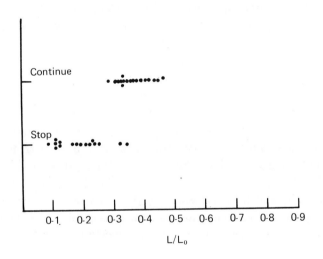

L/L₀

Fig. 6-8.   Continuation of beating in the proximal portion of the sperm tail as a function of its length following ruby laser microirradiation. Points indicate individual flagella: *L*, length of proximal portion; *Lo*, length of entire tail. (From S. Goldstein in *J. Exptl. Biol.*, **51**, 1969.)

Microirradiation of sperm has served also for the creation of a model of flagellum activity.   Goldstein (1969) used a ruby laser microbeam to produce localized damage along the flagellum of sperm of the sea urchin *Strongylocentrotus* and the starfish *Piaster*.   The purpose of his studies was to localize function and examine functional interactions along the length of the flagellum. Several observations relative to flagellum beat activity were made: (1) if the length of the tail between the irradiated point and the head was at least 25 percent of the total length of the flagellum, it would beat at least a few times after irradiation, but it would not beat at all if the unirradiated portion was shorter than 25 percent (Fig. 6-8); (2) beats already established beyond the irradiated point continued to propagate to the tip of the flagellum; (3) new bends did not develop past the irradiated point; and (4) irradiation within a bent region often completely eliminated the wave. In a later study Goldstein performed similar studies in the flagellum and cell body of a *Trypanasoid,* and demonstrated different responses (Fig. 6-9). He found that the portion of the flagellum between the cell body and the irradiated point beat from the base after irradiation, and that the piece of severed flagellum beat from either end for up to 10 cycles.

		Behaviour of flagellum	
Wave direction	Laser target	Between body and target	Between target and tip
Tip to base	Tip	Reversal of wave direction	–
	Base	–	Detached, independent propulsion; occasional reversal
	Middle	Reversal of wave direction	Detached, independent propulsion; occasional reversal
Base to tip	Tip	No reversal	–
	Base	–	Detached, independent propulsion; occasional reversal
	Middle	No reversal	Detached, independent propulsion; occasional reversal

Fig. 6-9. Summary of results of flagellum behavior in the trypanosoid *Crithidia* following ruby laser microbeam irradiation. (From S. Goldstein et al., in *J. Exptl. Biol.,* **53**, 1970.)

It appeared that in this organism there is not a specific region of the flagellum or cell body that is necessary for activity. Instead, Goldstein proposes that there are autonomous initiator sites along the flagellum. These particular studies are reminiscent of the early microbeam studies of Terni (1933), in which he irradiated the head, neck, and flagellum of sperm from the salamander *Geotriton.* He found that irradiations of the head or neck did not affect flagellum activity, and when the flagellum itself was irradiated, only the actual region exposed to the radiation ceased to beat. Flagella activity did continue on either side of the irradiated region.

Female sex cells also have been used in partial cell irradiation. Campbell and Inoúe (1965) studied the role of the centrioles in the maintenance of spindle birefringence by UV microirradiation of metaphase-arrested spindles in oocytes of *Pectinaria.*

The discussion so far in this chapter has dealt with studies of basic cellular processes employing microirradiation of gametes or pregametic cells. Similarly, basic cellular processes have been studied by microirradiation of early developing embryos. Some of Tchakhotine's earliest studies in the 1920s involved a study of the cell surface by UV microirradiation of unfertilized sea urchin (*Strongylocentrotus*) and bivalve mollusc (*Pholas*) eggs. Irradiation of a small surface area, followed by placement in hyper- or hypotonic medium resulted in marked swelling and distortion of the eggs. However, when similar experiments were performed following fertilization, the sensitivity of the egg surface was markedly changed. These were some of the earliest studies indicating that definite changes in membrane permeability and structure occurred as a direct result of fertilization. In addition, these early studies further emphasized the roles of the cell surface in osmoregulation and UV sensitivity. In a later study (1937) Tchakhotine demonstrated that it was possible to promote fusion of two embryos by UV microirradiation of a small portion of the surface of two sea urchin eggs and then gently compressing the eggs so that the irradiated regions were in contact. The embryos became fused and development proceeded for a short time.

Another basic cellular mechanism studied by microbeam irradiation of embryonic systems is cell division. Tchakhotine (1920) demonstrated that ultraviolet (2800 Å) irradiation of the nucleus in one blastomere of the two cell stage of the sea urchin *Strongylocentrotus* resulted in a significant delay in division in the irradiated cell. Irradiation of a similar volume of cytoplasm with twice as much radiation dose did not affect cell division. Zirkle (1932) similarly studied the delay and inhibition of cell division in the germination of the spores of the fern *Pteris,* using polonium alpha particles. Because the spore has an eccentrically located nucleus it was possible to selectively irradiate either the nucleus or cytoplasm. As in Tchakhotine's early studies there was a marked delay in division when the region of the cell containing the nucleus was irradiated, and no effect in division when the cytoplasm was irradiated.

There have been a large number of microbeam studies the principle goals of which have been a better understanding of the developing organism. As in most other areas of microbeam irradiation, the initial work was by Tchakhotine. In 1929 he demonstrated that it was possible to activate unfertilized sea urchin eggs by localized UV microirradiation (2800 Å) of the egg surface. These studies indicated that the results of the early studies of Loeb (1914) demonstrating that whole egg UV irradiation resulted in initiation of development, were due to a cell-surface effect. In addition, Tchakhotine's studies demonstrated that development could be initiated by affecting only a small portion of the egg surface, as in the case of normal fertilization. However, development in all cases was limited, usually only a few abnormal cleavages resulting. In later studies Tchakhotine perfected the microbeam size and optimized the radiation dose (1935, 1938). Substantially more advanced development was obtained by irradiating immature eggs of the mollusc *Pholas.*

Shortly after Tchakhotine's early studies, Vintemberger (1929) used X-ray partial cell irradiation on various portions of frog eggs either containing or lacking the nucleus. He demonstrated that most abnormalities occurred when the irradiated fraction contained the nucleus, thus, implicating this structure in embryonic radiosensitivity and suggesting, also, that it is the controlling structure in the developmental process.

A similar approach to the study of early invertebrate development was taken by Ulrich in the 1950s. In several studies on the development of the fruit fly *Drosophila,* he used an X-ray microirradiation device to expose various portions of the ellipsoid egg. Unlike vertebrate developing systems, in insects the early nuclear divisions occur inside of the cytoplasm of the egg, without a concomitant division of the cytoplasm. Using emergence of the larva from the egg capsule as an assay for radiation injury, Ulrich demonstrated an exponential dose mortality curve with an LD_{50} at about 500 r when the anterior nucleus-containing half of the egg was irradiated, and a sigmoidal curve with an LD_{50} at 90,000 r, when the posterior nonnucleus-containing half of the embryo was irradiated (Fig. 6-10).   These results further indicated that the nucleus is the primary site of radiosensitivity in the embryo, and that the early nuclei of *Drosophila* embryos are located in the anterior region of the embryo.   Because of the two different mortality curves, two different modes of killing were hypothesized.   However, both modes of killing apparently exhibited an oxygen effect; it was much more difficult to kill the embryos when irradiation was performed in a nitrogen enriched atmosphere.   A rather interesting result was the production of greater than 27% of hatching flies with abnormal abdomens when the middle $\frac{1}{5}$ of $\frac{1}{2}$ hr old eggs were irradiated, but when 4-5 day old embryos were irradiated, abnormal abdomens were obtained only when the bottom two-fifths of the embryos were irradiated.   These results suggested either a migration of specifically determined nuclei to the lower portion of the embryo by 4-5 days of development, or a similar migration and localization of a specific cytoplasmic morphogenetic constituent.

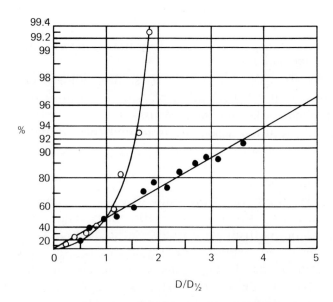

D/D½

Fig. 6-10. Percentage of *Drosophila* eggs killed following X-ray irradiation of either the nucleus-containing (•) or the nonnuclear half (o) of the egg. (From Ulrich, 1956.)

Remarkably similar results were obtained in eggs of the parasitic wasp *Habrobracon* when the anterior end of the large eggs (600 μ long and 150 microns wide) was irradiated with polychromatic UV radiation. An exponential mortality curve was obtained, and when the nonnucleus-containing portion of the embryo was irradiated, a sigmoid curve was obtained and the LD_{50} was attained with 15 times as much radiation dosage (Rogers and von Borstel, 1955).

Perhaps some of the most extensive series of studies on embryonic systems were performed by Kalthoff on the embryo of the insect *Smittia*. What makes these studies particularly significant is that they were directed at one of the most intriguing and fundamental concepts of developmental biology—pattern formation. One of the major tenets of pattern formation theory is that the formation of a spatial pattern is due to the release of different gene activities in an ordered spatial sequence. It is felt that the cytoplasm plays a critical role in controlling or triggering the activation of these genes. More specifically, it has been suggested that the cytoplasm of the egg, the ooplasm, is endowed during oogenesis with specific regional constituents that are responsible for releasing specific genes. Kalthoff (1971) points out that there are only a few results demonstrating the existence of prelocalized cytoplasmic differences based upon descriptive methods and that the numerous experimental studies are based on statistical results, rather than precisely controlled experiments. Furthermore, it is pointed out that individual treatment of each egg is very time consuming. "In order to establish functional correlations between constituents of the egg and pattern formation, a system should be used which reacts upon a simple bulk operation by formation, with high reliability, of an aberrant pattern."[1]

A system is described in the midge *Smittia* that "reacts to suitably adjusted UV-irradiation of the anterior pole region by producing, with 100% frequency, an aberrant pattern called 'double abdomen.' "[2] By lining eggs up in a groove under a coverslip against a slide, with the target region exposed to the radiation, it was possible to irradiate a number of eggs at the same time (Fig. 6-11). Using this procedure and a modified procedure for single eggs (Fig.

[1] K. Kalthoff, "Position of Targets and Period of Competence for UV-induction of the Malformation 'Double Abdomen' in the Egg of *Smittia* spec. (Diptera, Chironomidae)", *Wilhelm Roux' Archiv.*, 168 (1971), 64.

[2] *Ibid.*, p. 65.

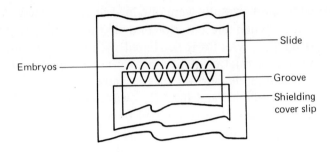

Fig. 6-11.   Schematic of Kalthoff's method for irradiation of a number of eggs. (From Kalthoff, 1971.)

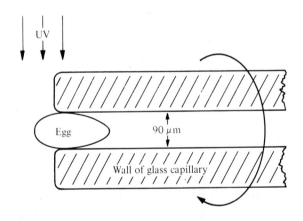

Fig. 6-12.   Schematical drawing of an egg sucked up in a glass capillary so that the anterior fourth can be irradiated with UV. This device has been employed for irradiation of the eggs from distinct sides and for rotation during irradiation. (From Kalthoff, 1971.)

6-12), Kalthoff performed experiments (1) varying UV dose and stage of development at irradiation, (2) varying dose and irradiated area, (3) testing UV absorption within the egg, (4) irradiating the anterior fourth from different sides, (5) rotating the eggs during irradiation, (6) irradiating central and peripheral regions within the anterior fourth of the eggs, and (7) determining the dose-response curve for irradiation of the anterior fourth at the two polar cell stage.   The results of these experiments, relative to the initial question of the localization of the targets for UV induction of double abdomens, are summarized in Fig. 6-13, which shows the

ap

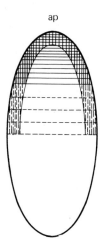

Fig. 6-13. Efficiency of irradiation of different areas, indicated by the density of hatching. The varying efficiency possibly represents a corresponding distribution in the concentration of effective targets or the morphogenetic efficiency of the targets in the irradiated state. *ap,* anterior pole. (From Kalthoff, 1971.)

efficiency of irradiation of different egg areas. The figure is based upon the following results: "(1) The extension of the irradiated area beyond the anterior half with constant UV dose scarcely increases the frequency of double abdomens at the expense of the normal larvae; (2) Double abdomens can be induced by irradiation of the first but not of the second fourth; (3) Irradiation of the first (most anterior) eighth yields more double abdomens than irradiation of the second eighth; (4) Neglecting some irregularities, the efficiency of irradiation decreases with increasing longitudinal extent of the irradiated anterior area."[3] Kalthoff stressed that irradiation of the anterior eighth is sufficient to inhibit the formation of such structures as the head, thorax, and anterior abdominal segments, which occupy more than *half* the egg. Furthermore the double abdomens were usually the same size. It is concluded that: (1) the effect on the irradiated targets is not confined to irradiated area; (2) no crucial egg regions exist where irradiation must occur in order to induce double abdomens; (3) the varying efficiency of irradiation of different egg areas represents a corresponding distribution in the concentration of targets or in the morphogenetic efficiency of targets in the irradiated area; (4) these targets are localized in the peripheral regions of the egg symmetrical to the longitudinal axis; and (5) since irradiation of cytoplasm *without* nuclei results in double abdomen production, the efficiency of double abdomen production does not

[3] *Ibid.,* p. 77.

change when the nuclei are in the irradiated ooplasm, proving that the targets are cytoplasmic and *not* nuclear.

Other developmental studies in insects involved UV microirradiation of various anlagen, and Richter (1966) studied eggs from the insect *Rhabelitis* by UV microirradiation after fertilization. Of particular interest in these studies were the cleavage planes of irradiated blastomeres.

Not to slight the plants, Zirkle (1932) caused abnormal development in the gametophytes of the fern *Pteris* by alpha particle bombardment of cytoplasm. The result was the formation of twin filaments which grew into double plates. This result could be explained if the irradiation caused the first plane of division to be perpendicular to its normal division plane. It is interesting to note that in this case, also, the developmental abnormality resulted from a cytoplasmic, rather than nuclear irradiation.

Despite the numerous studies on embryonic systems, one of the major problems encountered with UV microirradiation is penetrance of the beam. Because of the efficient absorption by cellular components, much of the radiation is absorbed by the first few cell layers. Consequently, microirradiation of underlying cells and tissues is often difficult.

Intense focused visible laser radiation has been employed by several investigators in the study of embryonic development. McKinnell (1969) used the classical ruby microbeam system to enucleate amphibian eggs prior to either fertilization with normal sperm or transplantation of a diploid nucleus. In the former case, development proceeded and embryos with the typical *haploid* syndrome were obtained. In the nuclear transplantation studies, development occurred, but a substantial number of embryos developed abnormal limbs. This result has not been adequately explained. In a more recent study Mims et al. (1971) used a ruby microbeam to irradiate the germinal crescent of developing chick embryos. In this preliminary study it was demonstrated that laser ablation of the primordial germ (sex) cells of the chick was possible without causing lethal damage to the embryo. Since classical extirpation studies usually resulted in embryo death, the laser technique should permit fruitful studies on the migration and development of these specialized cells. An earlier study by Barnes et al. (1965) involved ruby laser microirradiation of presomitic inducing regions in chick embryos. They used this approach to destroy various regions of the chick embryo that are involved in the formation of somites.

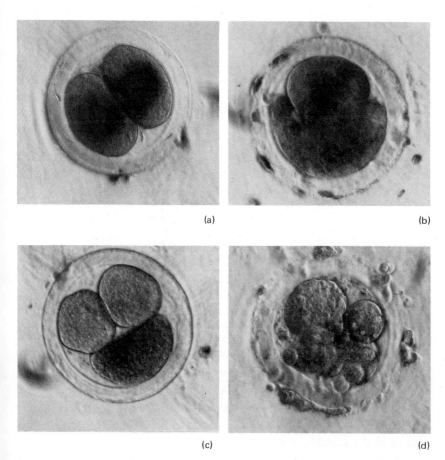

(a)   (b)

(c)   (d)

Fig. 6-14. Sequence of ova selected to show: (a) stained two-celled stage before lasing; (b) the collapse of the lased blastomere; (c) the surviving blastomere after one cleavage; (d) the resulting cells after four or more cleavages. (From Daniel and Takahashi, 1965.)

A rather interesting study on the irradiation of blastomeres of the two, four, and six cell stage of the rabbit embryos was conducted by Daniel and Takahashi in 1965. These authors found that when all but one cell of the various stages was destroyed by the laser microbeam (Fig. 6-14), the remaining cell continued to divide in culture. This is a rather significant finding in terms of rabbit blastomere totepotency (the ability of isolated cells to promote normal development). Another interesting feature of this study was the calibration of the energy density in the laser focused spot. By using the known melting temperatures of various metal alloys, the authors calculated the energy distribution curves of the focused laser spot (Fig. 6-15). This is probably the only laser study in which such precise information is available.

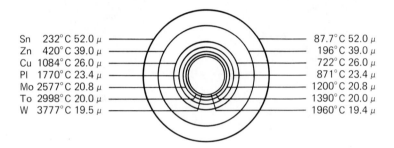

Sn 232°C 52.0 μ
Zn 420°C 39.0 μ
Cu 1084°C 26.0 μ
Pl 1770°C 23.4 μ
Mo 2577°C 20.8 μ
To 2998°C 20.0 μ
W 3777°C 19.5 μ

87.7°C 52.0 μ
196°C 39.0 μ
722°C 26.0 μ
871°C 23.4 μ
1200°C 20.8 μ
1390°C 20.0 μ
1960°C 19.4 μ

Fig. 6-15.   Temperature pattern of the focused laser beam determined from the melting points of various metals and the calculated temperatures of a stained blastomere at comparable distances from the focal point. (From Daniel and Takahashi, 1965.)

It is evident from the above discussion that all of the laser studies on embryos thus far have used the red ruby laser. This is due to the fact that it was the first laser system commercially available. Similar studies should be equally as possible with the green argon laser, and the tunable dye laser.

Another application of lasers in developmental studies is in plants. Though the number of studies is small, the potential for future investigations is great. Considerable work with the argon laser microbeam has already been done on the growth and differentiation of the green alga *Coleochaete* discussed earlier. Another plant, the moss *Protonemata,* has been studied by using the ruby laser to rupture the cell walls of all but one cell. In this way it was possbile to study the development of the single remaining apical or subapical cells (Leppard, 1964, from Zirkle's review). Saks et al. (1965) also have studied the growth of the plant *Nitella axilaris* by

Average increase in length per day (μ)	Number of Pulses of Irradiation							
	Nonirradiated		IX		3X		5X		7X	

Average increase in length per day (μ)	Nonirradiated		1	2*	1	2	1	2	1	2
First internodal	475	216	140	244	88	359	47	140	146	292
Second internodal	635	635	318	159	635	319	159	212	477	212
Total increase	1110	851	458	403	723	678	206	352	623	504
Percent decrease of irradiated cells			56		39		72		39	

*All number 2 internodal cells were subjected to 12.5 mJ from the incident beam with a beam diameter of 19.5 μ.

Fig. 6-16. Growth of internodal cells of *Nitella* following ruby laser microirradiation. (From Saks et al., 1965.)

irradiation of internodal cells with a ruby laser. They described a reduced growth rate of internodal cells when they were laser irradiated (Fig. 6-16). In addition it was demonstrated that the ruby microbeam could be used to punch very small holes in the cell wall of *Nitella,* thus permitting microinjection of oil droplets (Fig. 6-17). The authors point out that this kind of an approach would permit delicate microsurgery that has been extremely difficult in the past.

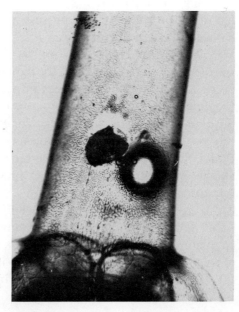

Fig. 6-17. Oil droplet injected into *Nitella* following puncture of cell wall with a ruby laser microbeam. (From Saks et al., 1965.)

Ultrastructure

SEVEN

One of the major goals of partial cell irradiation is to study cell function by selective alteration of a specific organelle or cell region. Implicit in this approach is a precise definition of structural changes resulting from the irradiation. This is particularly important when modification of a cellular process is attributed to radiation-induced damage to a target structure.

Numerous techniques of cytochemistry, light microscopy, histology, and radioautography have already been discussed. However, none of these techniques provide adequate information about the *fine-structural* changes produced by the microbeam irradiation. Studies with the electron microscope provided the first real look at the sub-light microscope changes produced by partial cell irradiation. Studies of this nature were first published by William Bloom at the University of Chicago in 1960 employing conventional UV microirradiation of chromosomes. Subsequent studies by Amy and Storb in Paris in 1965-1966 involved ruby laser irradiated mitochondria. Montgomery et al. in 1966 employed the electron microscope to analyse UV-irradiated chromosomes, and at about the same time, Moreno in Paris started publishing papers on conventional and laser UV-induced ultrastructural changes in various regions of the nucleus. In 1969 Griffin published an ultrastructural description of the changes produced in the plasmodium (*foot*) of the slime mold *Physarum* by ruby laser microirradiation. Most recently, Moreno and Salet (Paris) and Adkisson and Berns (U.S.A.) have described some of the ultrastructural changes produced in the mitochondria of contractile heart cells following irradiation with green laser (argon and neodymium) light.

The above mentioned group of studies is rather small considering the voluminous literature of partial cell irradiation. However, whenever one combines two rather sophisticated and expensive techniques such as electron microscopy and partial cell irradiation, the resultant number of studies is bound to be small initially. Also contributing to the small number of published studies is the technical difficulty of recovering and refinding selected irradiated cells, and then identifying the irradiated structure or region within the cell. These particular problems required extensive modification of the standardized EM procedures, and resulted in the development of a rather specialized EM technology for single cells.

A. METHODOLOGY OF SINGLE CELL RECOVERY

In their 1969 review, Moreno et al. succinctly stated three essential necessities of single cell recovery: (1) embed a monolayer of cells obtained from the surface of a slide or coverslip; (2) identify the irradiated cell, not only on the glass slide that serves as the object support during irradiation, but also in the plastic embedding medium; (3) make serial sections of the cell on a plane perpendicular or parallel to the axis of the irradiating beam. In addition to these requirements, one must also consider: (1) the difficulty in separating the cells from the glass or quartz coverslip; (2) recovering all or even most of the thin sections containing the sectioned cell; (3) relocating the cells on the electron microscope grid; (4) delineating between fixation artifacts and other distortions normally encountered in electron microscopy, and those effects caused by partial cell irradiation. It is no wonder that so few investigators have risked this approach.

In one of the first articles combining electron microscopy with partial cell irradiation, Bloom described the "Preparation of a selected cell for electron microscopy," (*J. Biophys. and Biochem.*

Cytol. 7:191, 1960). The cells were grown on quartz coverslips that were precoated with a thin film of evaporated carbon particles, which facilitated separation of cells from coverslip following embedding. After a cell was selected for study, a ring of about 250 μ was drawn around it on the coverslip with a diamond marking pen. In addition, careful diagrams and photographs were made of the cell and the surrounding neighboring cells. Fixation and dehydration were accomplished by floating the coverslip, cellside down, onto the appropriate solutions. Embedding was carried out by passing the coverslip with the cells from absolute alcohol into various 50:50 mixtures of solvent (ethylene dichloride) and alcohol, and finally to pure solvent before placing a couple of drops of solvent-methacrylate (embedding medium) over the cells. When the solvent evaporated (usually overnight) the cells were embedded in a thin film of methacrylate thus permitting the identification and precise relocation of the irradiated cell. The surrounding cells were cut away with a razor blade, and the remaining culture mounted on a lucite rod which fit into the chuck of the ultramicrotome. The culture was mounted on the tip of the rod by placing a small drop of viscous methacrylate directly over the target cell and attaching it to the rod.

After several days the culture plus coverglass was firmly attached to the lucite rod. By gently applying a small amount of dry ice to the coverslip, a fraction plan formed between the coverslip and the cells along the carbon film. The result was a clean separation of coverglass from cells which remained attached to the lucite rod. The culture was next trimmed for cutting using a dissecting microscope and razor blades mounted in a specifically designed guillotine-like holder that permits rapid and more accurate trimming of the block. The universal joint on the chuck of the ultramicrotome was replaced with a rigid chuck which held the lucite rod with the embedded cell on its tip. This modification assured that the cell was sectioned in the plane of the original surface of contact of the cell with the coverslip.

The above procedure and various modifications of it were used extensively by Bloom and his colleagues for electron microscopic interpretation of UV microirradiated cells.

A difficulty in applying the preceding technique to laser microbeam studies is the fact that carbon coated slides can not be employed because the carbon particles absorb the laser light. In their studies on ruby laser irradiated tissue culture cells, Storb et al. (1966) coated the coverslips with a thin film of formvar which served also as a fracture plane between coverslip and cells and was transparent to the laser beam. In later studies on the mitochondria of heart cells (Adkisson and Berns, unpublished) a thin film of silicone (*Siliclad*) was placed on the coverslip instead of formvar.

B.   CHROMOSOME PALING

In 1962 Bloom and Leider published a major paper describing the ultrastructural changes in UV-microirradiated chromosomes. Their extensive analysis employed both light microscope cyto-chemistry and electron microscopy. Salamander chromosomes were irradiated with approximately 10^{-1} ergs/μ^2 of heterochromatic UV light. At the site of irradiation the chromosomes exhibited the typical change of refractive index, thus appearing as a paled segment when viewed with dark field phase optics.

When assayed cytochemically these paled segments demon-strated a loss of DNA by virtue of negative Feulgen staining, and a reduced basic protein content as suggested by decreased staining by the Alfert-Geschwind procedure. In addition, Perry (1957) had previously demonstrated a reduced UV absorption at 2400, 2600, and 2800 Å. When similarly irradiated chromosomes were recovered with the electron microscope, striking ultrastructural changes were evident. It was immediately obvious that the type of fixation employed was most critical. For example, when standard fixation with OsO_4 was employed, no difference between the irradiated and unirradiated chromosome regions were detected. However, when neutral formalin-tyrode fixation was employed, definite differences between the irradiated and unirradiated chromosome regions were observed (Fig. 7-1). The unirradiated portions of the chromosomes appeared dark (electron dense) and homogeneous, whereas the irradiated segments appeared vacuolated and paler. This particular approach suggested that some component may have been removed from the chromosome thus leaving behind a vacuolated area. When heavy elemental staining with phosphotungstic acid was combined with the neutral formalin-tyrode fixation, more definite fine struc-tural details were evident. The normal unirradiated chromosome

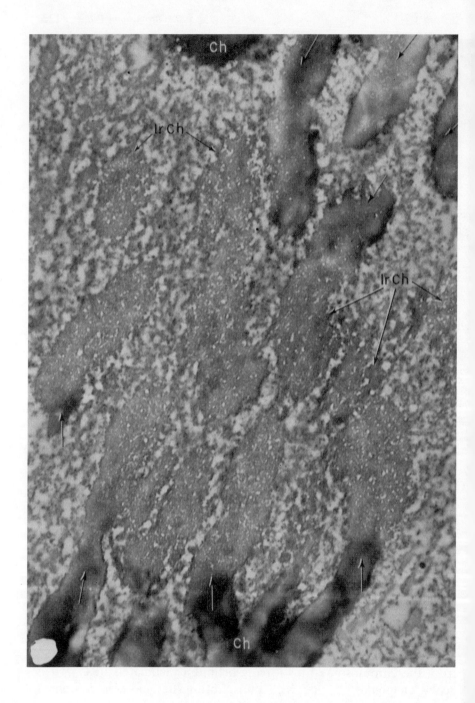

Fig. 7-1. Electron micrograph of chromosome exhibiting DNA steresis following UV microirradiation. The irradiated chromosome regions are indicated by the designation IrCh. Fixed with 10% neutral formalin; no stain. (From Bloom and Leider, 1962.)

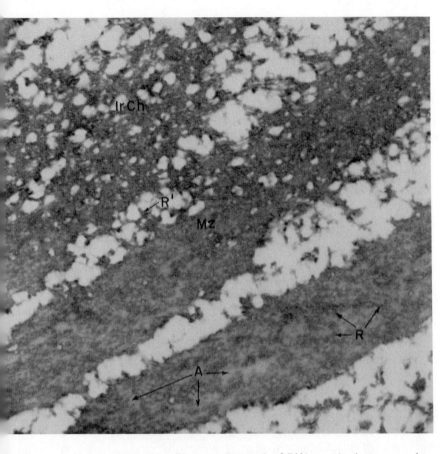

Fig. 7-2. Electron micrograph of DNA-steretic chromosome demonstrating the loss of component A. The irradiated region (IrCh) can be easily compared with the unirradiated chromosome which contains both components A and R. This cell was fixed in 10% neutral formalin-tyrode and stained with alcoholic phosphotungstic acid. (From Bloom and Leider, 1962.)

appeared to consist of two main components. The first component, R, was elongate branched strands of reticulum ranging in size from 40-300 Å. These strands, as seen in thin sections, were short segments of wavy, or bent, branching filaments which appeared to be part of a network of irregular meshes cut at random. Although these meshes of component R were frequently found next to each other, they were usually separated, singly or in groups, by a homogeneous or finely granular material, which is the second constituent, termed component A. It was component A that was lacking in the irradiated chromosome regions (Fig. 7-2). It can be seen that both components A and R are readily indentifiable in the unirradiated chromosome, but the irradiated chromosomes appear to have component R, plus the vacuoles, but little or no component A. From these studies the authors concluded that, indeed, component A contains DNA, and that this material disappears soon after UV microirradiation. The term DNA *steresis* was used to describe loss of this material. It was also suggested that perhaps component A contains some basic protein, because of the cytochemical and UV absorption data.

C. INTERPHASE NUCLEUS

The interphase nucleus has been studied ultrastructurally following UV microirradiation of the whole nuclei, nucleoli, or nucleoplasm by Montgomery et al. (1966), Moreno (1967), and Moreno et al. (1969a,b). In their studies with the flying spot TV technique Montgomery and colleagues exposed the entire nucleus or only the nucleolus of Chang liver cells in culture to heterochromatic UV light (0.18×10^{-17} erg/sec/cm^2) for periods from 2-24 hr. They found a segregation of nucleolar components resulting from either direct nucleolar irradiation or nuclear irradiation. In both cases segregation of nucleolar granules, loss of nucleolar organization, and clumping of nucleolar chromatin were observed. However, some differences were observed. Following nuclear irradiation the segregation resembled *dark caps* superimposed upon *light caps* at the nucleolar periphery. Following nucleolar irradiation, the light caps were not observed. The authors suggested that the latter nucleolar segregation resulted from a metabolic alteration in DNA metabolism, because of the close resemblance of the segregation to that observed when cells were treated with other inhibitors, such as actinomycin D, mitomycin C, etc.

In her 1967 paper, Moreno described nucleolar paling in interphase nuclei of KB cells that had been UV-microirradiated with 2312 Å light. Nucleolar irradiation resulted in a dispersion of the *perinucleolar* chromatin, and a further disaggregation of the nucleolar components. In addition to these direct effects on the irradiated nucleolus, a dispersion of chromatin clumps was produced throughout the nucleoplasm near the nuclear membrane and nucleoli.

Similar chromatin clumps were detected when the nucleus excluding the nucleoli were irradiated.

A later paper by Moreno and Vinzens (1969) presented a much more detailed examination of the ultrastructural effects of nuclear irradiation with the UV microbeam. In this paper both transverse and sagittal sections were made of the microirradiated cells, thus presenting a more comprehensive picture of the total cell damage produced by the microirradiation. It was found, for example, that the lesions had a transverse diameter of the same order of magnitude as the diameter of the irradiation beam, but, because of the distribution of energy in the focused spot, damage also occurred on either side of the focal plane. As in the previous report, chromatin clumps were observed following irradiation of whole nuclei, partial nuclei, or nucleoli. Dispersion of the perinucleolar chromatin and

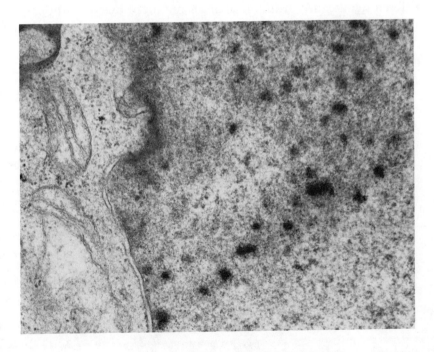

Fig. 7-3. Electron micrograph of nucleoplasm irradiated with a frequency quadrupled Q-switched neodymium laser (265 Å). Note the small electron-dense granules. 44,000X. (From Moreno, Salet, and Bessis in *C. R. Acad. Sci.,* 1969, Centrale des Revues Dunod-Gauthier-Villars.)

nucleolar components was also detected. It should be mentioned that specific damage to the cytoplasm and cell membrane above and below the irradiated nuclear region was detected also.

The studies discussed so far in this chapter have dealt with conventional UV microirradiation. There is one report in the literature of cells that were microirradiated in their nuclei with UV laser energy from a frequency quadrupled neodymium laser (2650 Å), (Morena et al., 1969 b). In these studies interphase nucleoplasm was irradiated with two different doses of light. At the high laser dose the irradiated area appeared as an electron-dense region with little discernible features. However, with the lower dose, even though no detectable change was seen with the light microscope, numerous small electron-dense granules (0.1 μ in diameter) were observed with the electron microscope (Fig. 7-3). These alterations differed significantly from the changes produced by conventional UV microirradiation. The authors speculated that the laser-induced changes result from thermal coagulation, and the conventional UV alterations result from a photochemical process.

D.   CYTOPLASM

Partial cytoplasmic irradiation generally has involved one of two approaches: (1) irradiation of a cytoplasmic area containing numerous organelles; (2) selective irradiation of single mitochondria.

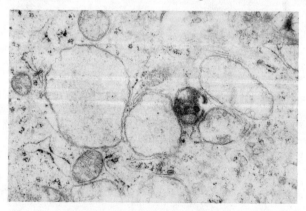

Fig. 7-4.   Conventional UV (2750 Å) lesion in the cytoplasm of KB liver cells. Note the swollen mitochondria and the dilated vacuoles. (From Moreno and Vinzens, 1969.)

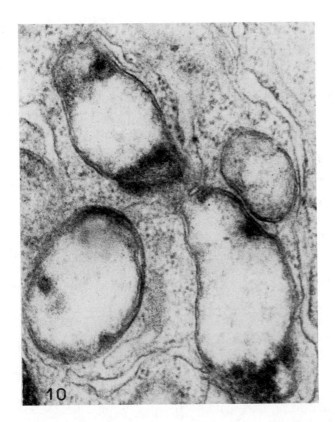

Fig. 7-5. Light primary damage to mitochondria following ruby laser irradiation (0.01 J incident energy) of Janus Green B (4 X 10^{-5}g/l) stained KB cells. Note electron dense material within the remnants of the mitochondria. (From Storb et al., 1966.)

In the first case, Moreno and Vinzens (1969) employed conventional 2750 Å UV light to irradiate a 12 μ^2 area of KB cells for periods of 30 sec to 2 min. Most obvious in the irradiated regions were swollen mitochondria and numerous vacuoles (Fig. 7-4). In addition, the endoplasmic reticulam were dilated. Little damage to the ribosomes was detected when irradiation was for less than 2 min, but with the longer irradiation exposure their number seemed to be reduced. In addition to these effects in the irradiated region, a general edema of the entire cell was detected. The nucleus, Golgi, and microfilaments did not appear to be affected. Also, as briefly mentioned in the previous section, damage to the cytoplasm and cell membranes above and below regions of nuclear irradiation exhibited effects similar to those described for direct cytoplasmic irradiation. The importance of this particular observation cannot be overemphasized, since numerous of the reported microbeam studies assume that the radiation damage is selective to the focal point of the beam.

In a similar study Storb et al. (1966) exposed the cytoplasm of tissue culture cells to ruby laser irradiation. In these studies the cells were first sensitized with the vital dye Janus Green B, and then a ruby laser beam was focused to a 5.8 μ spot in the cytoplasm. By varying both the concentration of the Janus Green stain, and the laser energy density, different degrees of cytoplasmic damage were produced. The damage was classified as either primary, within the area of irradiation, or secondary, outside of the area of irradiation but due to the irradiation. The primary damage was further subdivided into light, moderate, or heavy.

Light primary damage resulted from low level laser absorption (incident energy, 0.01 J; Janus Green B, 4×10^{-5}) and appeared to involve damage to only the mitochondria in the irradiated zone. Damaged mitochondria contained electron opaque material with portions of cristae remaining visible (Fig. 7-5).

Moderate primary damage resulted when a laser energy density of 0.02-0.092 J and a Janus Green B concentration of 6×10^{-5} was employed. Cells with these types of lesions also had damage restricted primarily to the mitochondria. The damage appeared as an electron-opaque region with numerous inner and outer mitochondrial membrane remnants attached. In addition to this primary damage, secondary damage limited to mitochondria outside the irradiation site was detected (Fig. 7-6). Typically, these damaged mitochondria were characterized by swelling, loss of cristae, lightening of the matrix, and the occurrence of small electron-opaque masses on the inner membrane. Surprisingly, these mitochondria morphologically recovered after 2-4 hr. However, there was strong evidence of irreversible damage at the molecular level. The cause of the secondary damage is not known, but it appeared to be a rather generalized effect resulting from the primary damage, rather than a specific effect due to a spreading of the laser beam from the site of primary damage.

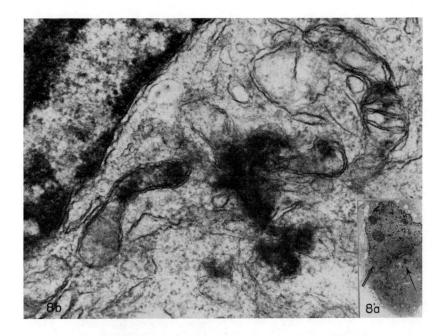

Fig. 7-6. Secondary damage (arrows) resulting from ruby laser (0.02-0.092 J) irradiation of Janus Green B stained (6×10^{-5}g/l) cells. Area of actual irradiation is not visible. Note swelling of mitochondria, loss of cristae, lightening of matrix, and the occurrence of small electron opaque masses on the inner membrane. (From Storb et al., 1966.)

Cells with heavy primary damage were grossly affected, both at the site of irradiation and outside this area. Virtually all organelles and the cell membranes were damaged or destroyed. These cells seldom remained viable for more than a couple of hours.

In all the experiments just described, the damage was attributed to absorption of the laser energy by the Janus Green B, which was bound selectively to the mitochondria. This result was substantiated by the fact that control cells not stained with Janus Green B, did not demonstrate any of the morphological alterations when exposed to similar levels of laser energy.

These studies are also particularly interesting because correlation can be made between ultrastructural changes, cell survival, and respiratory enzyme activity of the cells.

In a more recent study, Salet (1971) used the Green wavelength (5300 Å) of a frequency doubled neodymium laser to irradiate the mitochondrial rich cytoplasm of unstained heart cells in culture. In addition to seeing various changes in rate of contractility, specific ultrastructural changes in the mitochondria were detected. When a calculated energy density of 10^8 Wcm^2 was employed, no structural or beat frequency changes were observed; however, when an energy density of 10^9 Wcm^2 was used, contractility and ultrastructural changes were detected.   The major ultrastructural alterations were swollen mitochondria and damaged cristae (Fig. 7-7).

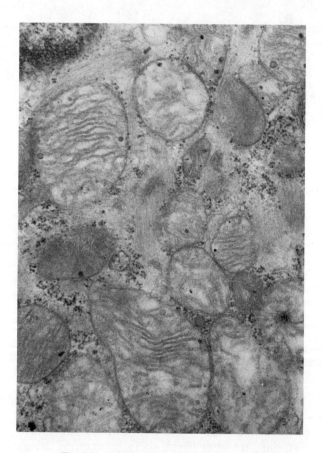

Fig. 7-7.   Mitochondria of a non-Janus Green stained rat myocardial cell irradiated with 10^9 W/cm^2 of 5300 Å laser light. Note swollen mitochondria and disorganized cristae. 68,000X. (From Salet, 1972.)

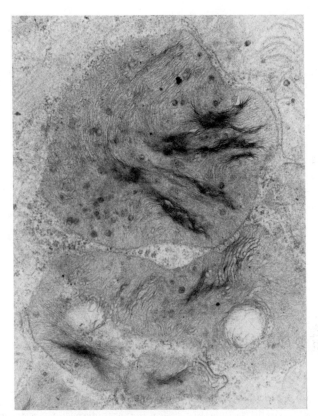

Fig. 7-8. Mitochondrion of Janus Green B stained cell (5.1^{-6} g/ml) irradiated with 5300 Å light 10^8 W/cm^2 from a neodymium laser. Note the electron opaque regions. (From Moreno, Salet, and Vinzens, 1973.)

In another more detailed ultrastructural analysis, Moreno and Salet and Vinzens (1973) compared the effects of green neodymium laser microirradiation and argon laser microirradiation on hearts cells, and KB liver cells. When the cells were vitally stained with Janus Green B (10^{-6} g/l), the laser light (10^8-10^9 Wcm^2) produced thermal lesions in the irradiated mitochondria (Fig. 7-8). When viewed with the electron microscope, electron opaque material was localized in the irradiated structures. Irradiation of nonvitally stained KB cells with either laser source did not produce mitochondrial alterations, but irradiation of unstained myocardial cell mitochondria with 10^8 Wcm^2 resulted in variation in beat frequency. The authors attribute this effect to specific absorption by the cytochromes.

Still another approach to the microirradiation of heart cells has been taken by Berns and colleagues. In these studies the green argon laser beam was focused directly into a single preselected mitochondrion. Various mitochondrial alterations have been described with the light microscope. These changes range from a slight phase paling of the irradiated mitochondrion to complete destruction of the target organelle.

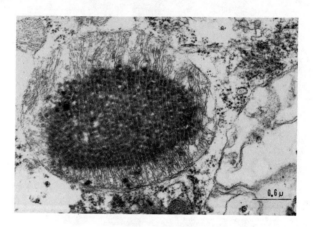

Fig. 7-9. Mitochondrion of rat myocardial cell irradiated with argon laser microbeam. Note the electron dense regions between the cristae. The cristae appear to traverse through this region and attach to the mitochondrial membranes.

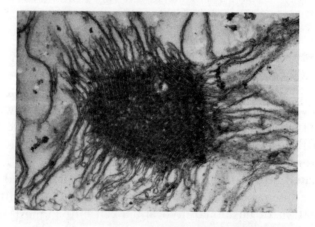

Fig. 7-10. Cristae pulled off mitochondrial membranes following argon laser microbeam irradiation. (From Berns, unpublished.)

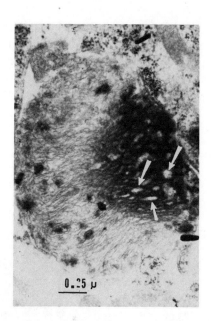

Fig. 7-11. Mitochondrion exhibiting areas apparently devoid of matrix material following argon laser microbeam irradiation. (From Berns, unpublished.)

Since the different mitochondrial alterations could be correlated with specific contractile responses, and the desired alterations could be produced using specific levels of laser energy or mitochondrial phase density, an ultrastructural characterization of the lesions seemed to be essential. Characterization of just the *least severe* (mitochondrial paling with small phase-dark spot in center) lesions demonstrated varied ultrastructural alterations. The most common ultrastructural change was the production of an electron dense area in the matrix between the cristae membranes. The cristae appeared not to be grossly affected (Fig. 7-9). However, other irradiated mitochondria appeared to have their cristae pulled off of the mitochondrial membranes (Fig. 7-10). Still others had disorganzied cristae, and some mitochondria had an electron dense area with regions apparently devoid of material (Fig. 7-11). The varied results suggested the changes with the light microscope were really quite variable in terms of ultrastructure. The reasons for this variability might be explained in terms of the absorption efficiencies of different mitochondria. The absorption coefficient of individual mitochondria could vary depending upon the thickness of the organelle, the concentration of the cristae per unit area, and the degree of oxidation/reduction of the various cytochromes. Even if the same amount of laser energy were focused into all the mitochondria, variability in any one of the above parameters could result in differences in laser absorption, thus resulting in a varied degree of damage.

The only other major ultrastructural analysis of laser micro-irradiated cells was performed on the slime mold *Physarum.* In these studies, Griffin et al. (1969) irradiated the cytoplasm of the *streaming plasmodium* using a ruby laser microbeam with various levels of energy. Ultrastructural alterations were detected consistently with energy levels greater than 600 J/cm². The typical

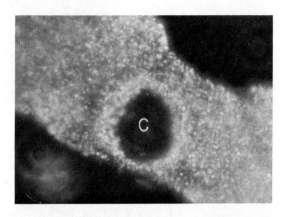

Fig. 7-12. Dark field light micrograph illustrating region of slime mold plasmodium fixed 45 sec following ruby laser microirradiation. Note absence of pigment granules in coagulum (c). Also note halo separating irradiated from unirradiated area. (From Griffin et al., 1969.)

light microscope lesion appeared as a coagulum that was separated from the unirradiated area by a halo (Fig. 7-12). When examined with the electron microscope, the halo region was shown to contain two membranes; the coagulum halo membrane, which most closely surrounds the irradiated region; and the protoplasmic halo membrane, which is apposed most closely to the unirradiated cytoplasm (Fig. 7-13). Upon irradiation the coagulum appeared to contract from the rest of the cytoplasm, leaving the halo space. This space was generally electron lucent, but contained some swollen and distorted mitochondria, pale nuclei, and infrequently clumped vesicles. In the central region of the coagulum, the cytoplasm appeared flocculent and dense and the membranes were distorted and broken. The authors concluded that many of the changes were caused by heating due to energy absorption by pigment granules. It was felt that many of the observed changes were due to secondary responses following the primary absorption of energy.

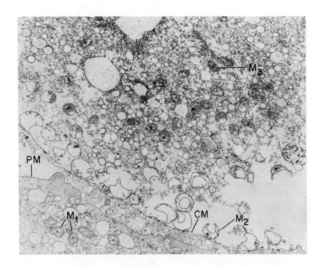

Fig. 7-13. Survey electron micrograph through irradiated plasmodium. Note space (bottom right), coagulum halo membrane (CM), protoplasmic halo membrane (PM), normal appearance of cytoplasm and mitochondria (M) outside the halo, the increased density of mitochondria (M$_3$) and cytoplasm in the center of the coagulum, and the swollen and distorted mitochondria (M$_2$) in the peripheral space left by contraction of the coagulum. (From Griffin et al., 1969.)

Recent electron microscope studies on laser microirradiated chromosomes have been performed by J. B. Rattner and M. W. Berns. Ultrastructural analysis has been conducted on the DNA lesion and protein lesion. Both lesion types appear to contain electron dense material in the chromosome area corresponding to the *paling* spot observed with the light microscope (Fig. 7-14). However, the nature of the electron dense material appeared different. The DNA lesion (which was produced following acridine orange treatment of the cells) contained small electron dense bodies 0.05μ in diameter scattered throughout the lesion area (Fig. 7-15). The

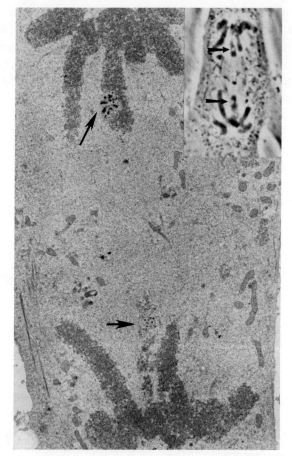

Fig. 7-14.   Low power electron micrograph of laser microirradiated anaphase cell. Arrows indicate lesion areas. The chromosomes were irradiated under the protein-damaging conditions (high laser power, no acridine orange). Inset is a phase micrograph of the cell immediately post-radiation.

electron dense material in the protein lesion appears as a number of contorted arrays of electron dense material, with a cross-sectional diameter of .08-.19μ, radiating from a large electron dense mass in the center of the lesion (Fig. 7-16). The damage produced under both conditions appeared to be limited to the irradiated chromosome region. Unirradiated chromosome areas and adjacent cytoplasm appeared unaffected.

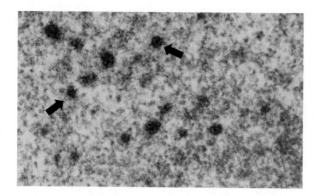

Fig. 7-15. High magnification of DNA lesion. Cells were treated with acridine orange (0.1 μg/ml, 5 minutes) and irradiated with moderate energy with the argon laser. Note the electron dense material 0.05μ in diameter.

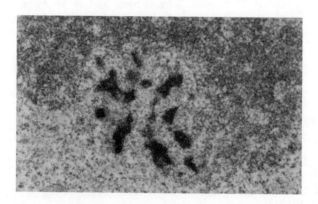

Fig. 7-16. High magnification of protein lesion (no acridine orange, high laser energy). Note the contorted arrays 0.08 - 0.19μ in diameter.

Bibliography

Adkisson, K., D. Baic, S. Burgott, W. K. Cheng, and M. W. Berns. Argon Laser Micro-irradiation of Mitochondria in Rat Myocardial Cells in Tissue Culture. IV. Ultrastructural and Cytochemical Analysis of Minimal Lesions. *J. Mol. and Cell. Cardiology.* In press, 1973.

Amenta, P. S. The Effects of Ultraviolet Microbeam Irradiation on the Eosinophil Granular Leukocytes of *Triturus viridescens. Anat. Record* 142:81-88, 1962.

Amy, R. L., and R. Storb. Selective Mitochondrial Damage by a Ruby Laser Microbeam: an Electron Microscopic Study. *Science* 150:756-757, 1965.

Amy, R. L., R. Storb, B. Fauconnier, and R. K. Wertz. Ruby Laser Microirradiation of Single Tissue Culture Cells Vitally Stained with Janus Green B. I. Effects Observed with the Phase Contrast Microscope. *Exptl. Cell Res.* 45361-373, 1967.

Bajer, A., and J. Mole-Bajer. U. V. Microbeam Irradiation of Chromosomes During Mitosis in Endosperm. *Exptl. Cell Res.* 25:251-267, 1961.

Barnes, F. S., K. Takahashi, and J. C. Daniel, Paper presented at the Nerem Meeting. Boston, Mass., November, 1965.

Basehoar, G., and M. W. Berns. Cloning of Rat Kangaroo (PTK$_2$) Cells Following Laser Microirradiation of Selected Mitotic Chromosomes. *Science* 179:1333-1334, 1973.

Berns, M. W. A Simple and Versatile Argon Laser Microbeam. *Exptl. Cell Res.* 65:470-473, 1971.

Berns, M. W. Partial Cell Irradiation with a Tunable Organic Dye Laser. *Nature* **240**:483-485, 1972.

Berns, M. W., and W. K. Cheng. Are Chromosome Secondary Constrictions Nucleolar Organizers: a Re-evaluation Using a Laser Microbeam. *Exptl. Cell Res.* **68**:185-192, 1971.

Berns, M. W., and W. K. Cheng. Mitotic Blockage Following Laser Microirradiation of Prophase Chromosomes. *Life Sciences* **11**:97-105, 1972.

Berns, M. W., W. K. Cheng, A. D. Floyd, and Y. Ohnuki. Chromosome Lesions Produced with an Argon Laser Microbeam Without Dye Sensitization. *Science* **171**:903-905, 1971.

Berns, M. W., and A. D. Floyd. Chromosome Microdissection by Laser: a Functional and Cytochemical Analysis. *Exptl. Cell Res.* **67**:305-310, 1971.

Berns, M. W., A. D. Floyd, K. Adkisson, W. K. Cheng, L. Moore, G. Hoover, K. Ustick, S. Burgott, and To Osial. Laster Microirradiation of the Nucleolar Organizer in Cells of the Rat Kangaroo *(Potorous tridactylis):* Reduction of Nucleolar Number and the Production of Micronucleoli. *Exptl. Cell Res.* **75**:424-432, 1972.

Berns, M. W., N. Gameleja, C. Duffy, R. Olson and D. E. Rounds, Argon Laser Microirradiation of Mitochondria in Rat Myocardial Cells in Tissue Culture. *J. Cell Physiol.* **76**:207-214, 1970.

Berns, M. W., D. C. L. Gross, and W. K. Cheng. Argon Laser Microirradiation of Mitochondria in Rat Myocardial cells in Tissue Culture, III. Irradiation of Multicellular Networks. *J. Molec. and Cell. Cardiol.* **4**:427-433, 1972.

Berns, M. W., D. C. L. Gross, W. K. Cheng, and D. Woodring. Argon Laser Microirradiation of Mitochondria in Rat Myocardial Cells In Tissue Culture. II. Correlation of Morphology and Function in Single Irradiated Cells. *J. Molec. and Cell. Cardiol.* **4**:71-83, 1972.

Berns, M. W., Y. Ohnuki, D. E. Rounds, and R. S. Olson. Modification of Nucleolar Expression Following Laser Micro-irradiation of Chromosomes. *Exptl. Cell Res.* **60**:133-138, 1970.

Berns, M. W., R. S. Olson, and D. E. Rounds. *In vitro* Production of Chromosomal Lesions Using an Argon Laser Microbeam. *Nature* **221**:74-75, 1969.

Berns, M. W., and D. E. Rounds. Cell Surgery by Laser. *Sci. Amer.* **222**:98-110, 1970.

Berns, M. W., D. E. Rounds, and R. S. Olson. Argon Laser Microirradiation of Nucleoli. *J. Cell Biol.* **43**:621-626, 1969.

Berns, M. W., and C. Salet. Laser Microbeams for Partial Cell Irradiation. *Int. Rev. Cytol.* **33**:131-155,1972.

Bessis, M., F. Gires, G. Mayer, and G. Nomarski. Irradiation des organites cellulaires a l'aide d'un laser a rubis. *C. R. Acad. Sci.* **225**:1010-1012, 1962.

Bessis, M., and G. Nomarski. Conditions de l'irradiation ultra-violette des organites cellulaires. *C. R. Acad. Sci.* **249** :768-776, 1959.

Bessis, M., and M. TerPogossian. Micropuncture of Cells by Means of a Laser Beam. *Ann. N. Y. Acad. Sci.* **122**:689-694, 1965.

Bloom, W. Preparation of a Selected Cell for Electron Microscopy. *J. Biophys. and Biochem. Cyt.* **7**:191-194, 1960.

Bloom, W., and R. J. Leider. Optical and Electron Microscopic Changes in Ultraviolet-irradiated Chromosome Segments. *J. Cell Biol.* **13**:269-301, 1962.

Bloom, W., R. E. Zirkle, and R. B. Uretz. Irradiation of Parts of Individual Cells. III. Effects of Chromosomal and Extrachromosomal Irradiation on Chromosome Movements. *Ann. N. Y. Acad. Sci.* **59**:503-513, 1955.

Campbell, D., and S. Inoúe. Reorganization of Spindle Components Following UV Microirradiation. *Biological Bulletin* **129**:401-402, 1965.

Caspersson, T., S. Farber, and G. E. Foley. Chemical Differentiation Along Metaphase Chromosomes. *Exptl. Cell. Res.* **49**:219-222, 1968.

Daniel, J. C., and K. Takahashi. Selective Laser Destruction of Rabbit Blastomeres and Continued Cleavage of Survivors *in vitro*. *Exptl. Cell Res.* **39**:475-479, 1965.

Davis, M. I., and C. L. Smith. The Irradiation of Individual Parts of Single Cells in Tissue Culture with a Microbeam of α-Particles. *Exptl. Cell Res.* **12**:15-34, 1957.

Deak, I., E. Sidebottom, and H. Harris. Further Experiments on the Role of the Nucleolus in the Expression of Structural Genes. *J. of Cell Science* **11**:379-391, 1972.

Dendy, P. P., and J. E. Cleaver. An Investigation of (a) Variation in Rate of DNA Synthesis During S-phase, (b) Effect of Ultra-violet Radiation on Rate of DNA-synthesis. *Int. J. Radiation Biol.* **8**:301-315, 1964.

Dendy, P. P., and C. L. Smith. Effects on DNA Synthesis of Localized Irradiation of Cells in Tissue Culture by (i) a U. V. Microbeam and (ii) an α-particle Microbeam. *Proc. Royal Soc.* (London) B160:328-344, 1964.

Fano, U. Principles of Radiological Physics, chap. 1 in *Radiation Biology*, Vol. 1, Part 1, A. Hollaender (ed.). McGraw Hill, New York, 1954.

Forer, A. Local Reduction of Soindle Fiber Birefrigence in Living *Nephrotoma suturalis* (Loew) Spermatocytes Induced by Ultraviolet Microbeam Irradiation. *J. Cell Biol.* **25**:95-117, 1965.

Forer, A. A Simple Conversion of Reflecting Lenses into Phase-Contrast Condensers for Ultraviolet Light Irradiations (Ultraviolet Microbeam Equipment, Ultraviolet Microscopes). *Exptl. Cell Res.* **43**:688-691, 1966.

Forer, A. Microbeam and Partial Cell Irradiation. *Proc. NATO Adv. Study Inst.*, Cannes, France, 1967.

Gaulden, M. E., and R. P. Perry. Influence of the Nucleolus on Mitosis as Revealed by UV Microbeam Irradiation. *Proc. Natl. Acad. Sci.* **44**:553-570, 1958.

Geyer-Duszynska, I. Experimental Research on Chromosome Elimination in Cecidomyidae (Diptera). *J. Exptl. Zool.* **141**:391-488, 1959.

Geyer-Duszynska, I. Spindle Disappearance and Chromosome Behavior After Partial-embryo Irradiation in Cecidomyidae (Diptera). *Chromosoma* **12**:233-247, 1961.

Glick, D. Cytochemical Analysis by Laser Microprobe-emission Spectroscopy. *Ann. N. Y. Acad. Sci.* **157**:265-274, 1969.

Glubrecht, H. Ein Monochromator fur die UV-Mikrospektroskopie. *Naturwissenschaften* **45**:33-35, 1958.

Glubrecht, H. Strahlenschaden bei der UV-Mikrospektroskopie. *Acta histochemica Bd.* **9**:195-199, 1960.

Glubrecht, H. Ultraviolettmikrospektrometrische Untersuchung partiell bestrahlter Pflanzen zellen. *Biophysik* **1**:78-86, 1963.

Goldstein, S. F. Irradiation of Sperm Tails by Laser Microbeam. *J. Exptl. Biol.* **51**:431-441, 1969.

Goldstein, S. F., M. E. J. Holwill, and N. R. Silvester. The Effects of Laser Microbeam Irradiation on the Flagellum of *Crithidia* (Strigomonas) *oncopelti. J. Exptl. Biol.* **53**: 401-409, 1970.

Griffin, J. L., M. N. Stein, and R. E. Stowell. Laser Microscope Irradiation of *Physarum polycephalum*: Dynamic and Ultrastructural Effects. *J. Cell Biol.* **40**:108-119, 1969.

Harris, H. *Nucleus and Cytoplasm.* Oxford University, London, England, 1970.

Inoúe, S., and H. Sato. Cell Motility by Labile Association of Molecules. The Nature of Mitotic Spindle Fibers and Their Role in Chromosome Movement. *J. Gen. Physiol.* **50**:259-288, 1967.

Izutsu, K. Effects of Ultraviolet Microbeam Irradiation Upon Division in Grasshopper Spermatocytes. I. Results of Irradiation During Prophase and Prometaphase. I. *Mie. Med. J.* **11**:199-212, 1961.

Jagger, J., D. M. Prescott, and M. E. Gaulden. A UV Microbeam Study of the Roles of Nucleus and Cytoplasm in Division Delay, Killing, and Photoreactivation of *Amoeba proteus. Exptl. Cell Res.* **58**:35-54, 1969.

Kalthoff, K. Position of Targets and Period of Competence for UV-induction of the Malformation "Double Abdomen" in the Egg of *Smittia* spec. (Diptera, Chironomidae). *Wilhelm Roux' Archiv.* **168**:63-84, 1971.

Lacalli, T. C., and A. B. Acton. An Inexpensive Laser Microbeam. *Trans. Amer. Micros. Soc.* **91**:236-238, 1972.

Leppard, G. G. M. A. Thesis, University of Saskatchewan, Canada, 1964.

Loeb, J. Activation of the Unfertilized Egg by Ultraviolet Rays. *Science* **40**:680-681, 1914.

McKinnell, R., M. F. Mims, and L. A. Reed. Laser Ablation of Maternal Chromosomes in Eggs of *Rana pipiens*. *Z. Zell.* **93**:30-35, 1969.

Mims, M. F., and R. McKinnell. Laser Irradiation of the Chick Embryo Germinal Crescent. *J. Embryol. Exp. Morph.* **26**:31-36, 1971.

Montgomery, P. O'B., and L. L. Hundley. UV Microbeam Irradiation of the Nucleoli of Living Cells. *Exptl. Cell Res.* **24**:1-5, 1961.

Montgomery, P. O'B., R. C. Reynolds, and J. E. Cook. Nucleolar "Caps" Induced by Flying Spot Ultraviolet Nuclear Irradiation. *Am. J. Pathol.* **49**:555-567, 1966.

Moore, L. B., and M. W. Berns. Microdissection of Actinomycin D Segregated Nucleoli with Laser. *In Vitro.* **8**:403, 1973.

Moreno, G. Etude au microscope electronique de l'irradiation ultraviolette localisee au noyau, au nucleole. *Compt. rend. soc. biol.* **161**:1866-1868, 1967.

Moreno, G. Effects of UV Microirradiation on Different Parts of the Cell. II. Cytological Observation and Unscheduled DNA Synthesis After Partial Nuclear Irradiation. *Exptl. Cell Res.* **65**:129-139, 1971.

Moreno, G., M.Lutz, and M. Bessis. Partial Cell Irradiation by Ultraviolet and Visible Light: Conventional and Laser Sources. *Int. Rev. Exp. Path.* **7**:99-137, 1969.

Moreno, G., C. Salet, and M. Bessis. Micro-irradiation de noyaux cellulaires par rayonnement laser ultraviolet. *C. R. Acad. Sci.* **269**:781-782, 1969b.

Moreno, G., C. Salet, and F. Vinzens. Etude en microscopie electronique des mitochondries de cellules en culture de tissus apres micro-irradiation par laser. *J. de Microscopie* **16**:269-278, 1973.

Moreno, G., and F. Vinzens. Effets de la micro-irradiation ultra-violette sur differentes parties de la cellule. I. Etude en microscopie electronique sur coupes en series transversales et sagittales. *Exptl. Cell Res.* **56**:75-83, 1969a.

Munro, T. R. Advances in Radiobiology. Proceedings 5th International Conference on Radiobiology, p. 108, G. De Hevesy et al., (eds.). Oliver and Boyd, Edinburgh and London, 1957.

Nakanishi, Y. H., and H. Kato. Unusual Movement of the Daughter Chromosome Group in Telephasic Cells Following the Exposure to Ultraviolet Microbeam Irradiation. *Cytologia* **30**:213-221. 1965.

Nakanishi, Y. H., and S. Makino. Cytogenetics of Cells in Culture. *Symp. Intern. Soc. Cell Biol.*, Vol. 3, 59-61, R. J. C. Harris (ed.). Academic Press, New York, 1964.

Ohnuki, Y., M. W. Berns, D. E. Rounds, and R. S. Olson. Laser Microbeam Irradiation of the Juxtanucleolar Region of Prophase Nucleolar Chromosomes. *Exptl. Cell Res. 71*:132-144, 1972.

Perry, R. P. Changes in the Ultraviolet Absorption Spectrum of Parts of Living Cells Following Irradiation with an Ultraviolet Microbeam. *Exptl. Cell Res.* **12**:546-559, 1957.

Perry, R. P., A. Hell, and M. Errera. The Role of the Nucleolus in Ribonucleic Acid—and Protein Synthesis. I. Incorporation of-Cytidine into Normal and Nucleolar Inactivated HeLa Cells. *Biochim. Biophys. Acta* **49**:47-57, 1961.

Petrova, J. Different effects of α-rays on Nucleus and Cytoplasm. *Botan. Centr. Beih.* **A61**:399-430, 1942.

Phillips, S. Repopulation of the Postmitotic Nucleolus by Preformed RNA. *J. Cell Biol.* **53**:611-633, 1972.

Pohlit, W. Eine Elecktronenkanone fur biophysikalische Untersunchungen. *Strahlentherapie* **103**:593-597, 1957.

Richter, I. E. Mikrozeitrafferuntersuchung zur UV-bedingten Teilungsverzogerung des *Rhabditis*-Eies. *Protoplasma* **62**:237-245, 1966.

Rogers, R. W., and R. C. von Borstel. Paper presented before *Am. Assoc. Advance. Sci.*, Atlanta, 1955.

Rustad, R. C. A Simple UV-microbeam for Partial Cell Irradiation. *Experientia* **24**:974-975, 1968.

Sakharov, V. N., and L. N. Voronkova. Consequences of UV-microbeam Irradiation of the Nucleoli of Living Cell. *Genetika* **6**:144-148, 1966.

Saks, N. M., and C. A. Roth. Ruby Laser as a Microsurgical Instrument. *Science* **141**:46-47, 1963.

Saks, N. M., R. Zuzolo, and M. J. Kopac. Microsurgery of Living Cells by Ruby Laser Irradiation. *Ann. N. Y. Acad. Sci.* **122**:695-712, 1965.

Salet, C. Acceleration par micro-irradiation laser du rhythme de contraction de cellules cardiaques en culture. *C. R. Acad. Sci. Paris* **272**:2584-2586, 1971.

Salet, C. A Study of Beating Frequency of a Single Myocardial Cell. I. Q-switched Laser Microirradiation of Mitochondria. *Exptl. Cell Res.* **73**:360-366, 1972.

Salet, C., and F. Vinzens. In press, 1973.

Sato, H., and K. J. Muller. An Analysis of Living Squid Sperm Head Fine Structure Through Polarized UV Microbeam Irradiation. *Biol. Bull.* 131:404-405, 1966.

Seidel, F., and C. Bucholtz. Die Dosisleistung bei Durchstrahlung biologischer Objekte mit nadelformigen Bundeln von Rontgenstrahlen. Naturwissenschaften 47:260-261, 1960.

Shimomura, T., S. Naruse, and S. Takeda. Stomatogenesis and Cell Division in *Euplotes* Inhibited by Ultra-violet Microbeam Irradiation. *Nature* 215:91-92, 1967.

Skreb, Y., and N. Skreb. Effects des rayons U. V. sur la respiration des fragments d'amibes. *Biochim. Biophys.* 39:540-541, 1960.

Smith, C. L. Effect of α-particle and X-ray Irradiation on DNA Synthesis in Tissue Cultures. *Proc. Roy. Soc. Biol. Sci.* 154:557-570, 1961.

Smith, C. L. Microbeam and Partial Cell Irradiation. *Int. Rev. Cytol.* 16:133-153, 1964.

Smith, K. C., and P. C. Hanawalt. *Molecular Photobiology—Inactivation and Recovery.* Academic Press, New York, 1969.

Storb, R., R. L. Amy, R. K. Wertz, B. Fauconnier, and M. Bessis. An Electron Microscope Study of Vitally Stained Single Cells Irradiated with a Ruby Laser Microbeam. *J. Cell Biol.* 31:11-29, 1966.

Tchakhotine, S. Die mikrikopische Strahlenstrich methode, eine Zelloperationsmethode. *Biol. Zentralbl.* 32:623, 1912.

Tchakhotine, S. Action localisee des rayons ultraviolets sur le noyau de l'oeuf d'Oursin, par radiopuncture microscopie. *Compt. rend. soc. biol.* 83:1593, 1920.

Tchakhotine, S. Les changements de la permeabilité de l'oeuf d'Oursin, localises experimentalement. *Compt. rend. soc. biol.* 84:464, 1921.

Tchakhotine, S. Attivazione dell'uovo di riccio di mare per mezzo della microraggiopuntura. *Boll. soc. ital. biol. sper.* 4:475-479, 1929.

Tchakhotine, S. Die Mikrostrahlstichmethode und andere Methoden des zytologischen Experimentes. *Handbuch biol. Arbeitsmethoden Abt.* V 10:877, 1935.

Tchakhotine, S. L'effet d'arret de la fonction de la vacuole pulsatile de la Paramecie par micropuncture ultraviolette. *Compt. rend. soc. biol.* 120:782-784, 1935.

Tchakhotine, S. Floculation localisée des colloides dans le cellule par la micropuncture ultraviolette. *Compt. rend.* 200:2036-2038, 1935.

Tchakhotine, S. Recherches physiologiques sur les Protozoaires, faites au moyen de la micropuncture ultraviolette. *Compt. rend.* 200:2217-2219, 1935.

Tchakhotine, S. Irradiation localisée du myonème du pedoncule des Vorticelles par micropuncture untraviolette. *Compt. rend.* **202**:1114, 1936.

Tchakhotine, S. La fonction du stigma chez le Flagelle Euglena, etudiée au moyen de la micropuncture ultraviolette. *Compt. rend. soc. biol.* **121**:1162-1165, 1936.

Tchakhotine, S. Das zytologische Mikroexperiment. (Untersuchungen an isolierten Zellen mit der Mikrostrahlstichmethode.) *Arch. exptl. Zellforsch. Gewebezucht.* **19**:498-506, 1937.

Tchakhotine, S. Parthenogenese experimentale de l'oeuf de la Pholade par micropuncture ultraviolette, aboutissant a une larve vivante. *Compt. rend.* **206**:377-379, 1938,

Terni, T. Microdissection et U. V. microradiopiqure des spermatozoides. *C. R. l'Assoc. Anat.* **28**:651, 1933.

Ulrich, H. The Results of Partial X-radiation of Drosophila Eggs. *Biol, Zentr.* **70**:274, 1951a.

Ulrich, H. Rontgenteilbestrahlung von Drosophila-Eiern. *Naturwissenschaften* **38**:121, 1951b.

Ulrich, H. A. Comparison of X-ray Effects on Nuclei and Plasma of Drosophila Eggs. *Biol. Zentr.* **74**:498, 1955a.

Ulrich, H. Die Bedeutung von kern und Plasma bei der Abtotung des Drosophila-Eiesdurch Rontgenstrahlen. *Naturwissenschaften* **42**:468, 1955b.

Ulrich, H. Die Strahlenempfindlichkeit von Zellkern und Plasma und die indirekte mutagene Wirkung der Strahlen. *Verhandl. Deut. Zool. Ges. Hamburg* **20**:150-182, 1956.

Uretz, R. B., W. Bloom, and R. E. Zirkle. Irradiation of Parts of Individual Cells. II. Effects of an Ultraviolet Microbeam Focused on Parts of Chromosomes. *Science* **120**:197-199, 1954.

Vintemberger, P. Sur une technique permettant d'irradier, dans des conditions de grande précision, une fraction determinée d'une cellule volumineuse comme l'oeuf de Grenouille rousse. *Compt. rend. soc. biol.* **102**:1050-1052, 1929.

Vintemberger, P. Sur les resultats de l'application d'une tres forte dose de rayons X a l'hemispher inferieur, anuclee de l'oeuf de Grenouille rousse. *Compt. rend. soc. biol.* **102**:1053-1055, 1929.

Vintemberger, P. Sur les resultats de l'application d'une tres forte dose de rayons X a diverses regions de l'oeuf de Grenouille rousse. *Compt. rend. soc. biol.* **102**:1055-1057, 1929.

Wada, B., and K. Izutsu. Effects of Ultraviolet Microbeam Irradiation on Mitosis Studied in *Tradescantia* Cells *in vivo*. *Cytologia* **26**:480-491, 1961a.

Wada, B., and K. Izutsu. Effects of Ultraviolet Microbeam Irradiations on Plant Mitotic Cells. *Saibo Kagaku Shinpojiumu* **11**:105-111, 1961b.

Wertz, R. K., R. Storb, and R. L. Amy. Ruby Laser Micro-irradiation of Single Tissue Culture Cells Vitally Stained with Janus Green B. III. Effects on the Incorporation of an RNA Pyrimidine Precursor. *Exptl. Cell Res.* **45**:61-71, 1967.

Wise, B. N. Effects of Ultraviolet Microbeam Irradiation on Morphogenesis in *Euplotes*. *J. Exptl, Zool.* **159**:241-268, 1965.

Withrow, R. B., and A. P. Withrow. Generation, Control, and Measurement of Visible and Near-Visible Radiant Energy, Chap. 3 in *Radiation Biology*, Vol. III, A. Hollaender (ed.). McGraw Hill, New York, 1954.

Zirkle, R. E. Some Effects of Alpha Radiation Upon Plant Cells. *J. Cell. Comp. Physiol.* **2**:251-274, 1932.

Zirkle, R. E. Partial Cell Irradiation. *Adv. Biol. Med. Phys.* **5**:103-146, 1957.

Zirkle, R. E. Ultraviolet-microbeam Irradiation of Newt-cell Cytoplasm: Spindle Destruction, False Anaphase, and Delay of True Anaphase. *Radiation Research* **41**:516-537, 1970.

Zirkle, R. E., and R. B. Uretz. Action Spectrum for Paling (Decrease in Refractive Index) of Ultraviolet-irradiated Chromosome Segments. *Proc. N. A. S.* **49**:45-53, 1963.

INDEX